Science and Fiction

More information about this series at http://www.springer.com/series/11657

Science and Fiction – A Springer Series

This collection of entertaining and thought-provoking books will appeal equally to science buffs, scientists and science-fiction fans. It was born out of the recognition that scientific discovery and the creation of plausible fictional scenarios are often two sides of the same coin. Each relies on an understanding of the way the world works, coupled with the imaginative ability to invent new or alternative explanations - and even other worlds. Authored by practicing scientists as well as writers of hard science fiction, these books explore and exploit the borderlands between accepted science and its fictional counterpart. Uncovering mutual influences, promoting fruitful interaction, narrating and analyzing fictional scenarios, together they serve as a reaction vessel for inspired new ideas in science, technology, and beyond.

Whether fiction, fact, or forever undecidable: the Springer Series "Science and Fiction" intends to go where no one has gone before!

Its largely non-technical books take several different approaches. Journey with their authors as they

- Indulge in science speculation – describing intriguing, plausible yet unproven ideas;
- Exploit science fiction for educational purposes and as a means of promoting critical thinking;
- Explore the interplay of science and science fiction – throughout the history of the genre and looking ahead;
- Delve into related topics including, but not limited to: science as a creative process, the limits of science, interplay of literature and knowledge;
- Tell fictional short stories built around well-defined scientific ideas, with a supplement summarizing the science underlying the plot.

Readers can look forward to a broad range of topics, as intriguing as they are important. Here just a few by way of illustration:

- Time travel, superluminal travel, wormholes, teleportation
- Extraterrestrial intelligence and alien civilizations
- Artificial intelligence, planetary brains, the universe as a computer, simulated worlds
- Non-anthropocentric viewpoints
- Synthetic biology, genetic engineering, developing nanotechnologies
- Eco/infrastructure/meteorite-impact disaster scenarios
- Future scenarios, transhumanism, posthumanism, intelligence explosion
- Virtual worlds, cyberspace dramas
- Consciousness and mind manipulation

Michael Carroll

Europa's Lost Expedition

A Scientific Novel

Michael Carroll
Littleton, CO, USA

ISSN 2197-1188   ISSN 2197-1196   (electronic)
Science and Fiction
ISBN 978-3-319-43158-1   ISBN 978-3-319-43159-8   (eBook)
DOI 10.1007/978-3-319-43159-8

Library of Congress Control Number: 2016947957

Printed on acid-free paper

Cover illustration: Painting by Michael Carroll; rover inset by Aldo Spadoni

This Springer imprint is published by Springer Nature
The registered company is Springer International Publishing AG Switzerland

My best friend and wife, Caroline, got me into mysteries in the first place. She had read every Agatha Christie story by the age of fifteen. My darling, life is full of wonderful mystery; this mystery is for you!

"Is it that by its indefiniteness it shadows forth the heartless voids and immensities of the universe, and thus stabs us from behind with the thought of annihilation, when beholding the white depths of the milky way?"

Moby Dick by Herman Melville

I have been a multitude of shapes,
Before I assumed a consistent form.
I have been a sword, narrow, variegated,
I have been a tear in the air,
I have been the faintest of stars.
I have been a word among letters,
I have been a book in the origin.

Sixth century poem by Taliesin

Acknowledgments

To anyone even vaguely familiar with murder mysteries, it will quickly become evident that this story has its roots in Agatha Christie's *And Then There Were None* (Section V of Chapter 7 is a specific homage). The classical mystery structure, her characterizations, and her sometimes twisted logic make her the Grande Dame of the genre; this book is a respectful nod to her. My thanks to Paul Schenk of the Lunar and Planetary Institute for help with Europa outpost sites. Special commendation goes to Robert Pappalardo for insights into moon innards and Jovian radiation, as well as serving as my "first reader" for the science section at the back. The first readings for a book like this are critical, and those were carried out by my fellow space artist Marilyn Flynn and my forever-loving and patience-of-Job wife Caroline. Caroline gets extra kudos for actually living through the process in the same house, not an easy thing to do.

Contents

Part I

The Novel

1

Ice Tomb

I

The Europa science outpost at Taliesin impressed even the most seasoned pro. Its web of habitats spread out from a central hub across acres of frozen wilderness. At the far end, connected by a single snakelike tunnel, stood The Dome. Melded to the ground with a slurry of water-turned-rock-hard-ice, the inflated balloon rose above a variety of crates, computers, storage units, and scaffolding, all within its pressurized embrace. A separate nuclear power plant fed the Dome's hungry appetite for electricity. Pumps inflated, lights illuminated, heaters warmed winter to—if not summer—at least late autumn (ambient temperatures were held to below freezing so as not to damage the ground underneath). And within this cacophony scampered a dozen engineers and scientists, biologists and core-drilling experts, bundled in arctic parkas, hoping to get at the ocean many kilometers below the glacial surface.

Joel Snelling gazed up at the drill's crane, a prehistoric monster of tubes and struts, hovering above the cavernous hole in the ice. He closed his eyes and forced in a breath of arctic ambience. The sharp bite of chilled air mingled with the smell of lubricants, the ozone of electricity, the sterile plastics. A lot was going on here.

Joel studied the apparatus with admiration. It had been a long haul. Simply getting a deep-drilling rig out this far was just short of miraculous. Of course it had been tried before. The first time, before the war, the attempt resulted in the loss of an entire expedition to Europa's southern hemisphere. The globe-spanning war had taken its toll on cities and people and infrastructure and art, but it had also taken a bite out of science. The entire Earth had devolved into

© Springer International Publishing Switzerland 2017
M. Carroll, *Europa's Lost Expedition*, Science and Fiction,
DOI 10.1007/978-3-319-43159-8_1

another Dark Age for years, and any resources that had been destined for the sciences went to bombs and pulse weapons. The names echoed in Joel's mind every time he thought of it: the Eastern Alliance, the Western Axis, the Seven Sisters and their death camps. Evil times.

But the world was recovering. The dust was settling and people had begun to heal and dream again. Here, beneath this dome, was a second chance, a chance to succeed where the lost expedition had failed. They kept improving the engineering, kept figuring out the problems and smoothing out the wrinkles. This time would be better.

The crane did a slow pirouette before him, a dancer showing off her graceful moves. But the metallic ballerina seemed off her footing. Abruptly, the crane's crossbeam dipped slightly.

"That doesn't look right," one of the techs called from a perch above the scaffolding.

A loud bang echoed against the walls of the dome, followed quickly by the sounds of grinding metal and tearing composites. Wires and cables, like so many tendons, shredded before Joel's eyes. This could only mean disaster. No time for the safety protocols. He had to make it to the door.

As the scaffolding and crane began to teeter above the hole, the tech yelled to him. "Hey, man, grab those cables so they don't fall in the drink. We can still save this thing. Joel!"

Joel had made it to the exit, amidst the shrieks of two engineers who had been dragged into the hole by the falling debris.

"Joel!" the tech called. But behind the man, Joel could see a spider web of fractures in the dome. The crane, rent from its base, had swung against it. The banshee scream of the vacuum raced through the chamber. His ears popped. Boxes and equipment migrated toward the opening. Darlene Esposito, one of Joel's favorite researchers, disappeared into the void through the gash in the dome, like Dorothy in her tornado. But for Darlene, there would be no Oz, only a deadly, radiation-soaked freeze under the stare of Jupiter.

"Joel!" He could barely hear the engineer's voice in the maelstrom. It was far too late to turn back. He had only seconds to open the hatch before the pressure dropped to a point where the door wouldn't open. He looked around for someone close enough that he might help, but no one was near. Another tech near the scaffolding held on desperately, trying to keep the base from flying apart. What if he helped her? What if their combined strength could turn the tide? But the door beckoned, and besides, she was a lost cause, wasn't she?

He turned to the hatch and pulled it open. He could hear the alarms now, bellowing down the corridor. He slammed the hatch behind him. As the wind died down, he heard the drone of the robotic voice:

Decompression alert, access two. Decompression alert, access two.

The long hallway would be sealed off any second. He ran as quickly as his bulky coat would allow. In the 1/8 gravity, his feet scrambled to find purchase. The tunnel's floor overlay the irregular ice beneath, its walls waving in the moving air. He turned a last corner just as he heard the door ahead of him clang shut. He bashed into it and pressed his face to the glass.

"Please!" he yelled to whoever was on the other side of the fogged glass. "It's still safe—let me in!"

He thought of those he had left behind. He thought about the tech clinging to the rig's base, and of Darlene Esposito. Sweet Darlene.

The hatch thumped. Two sets of arms pulled Joel into the light, into the air, into life. He hoped the closed-circuit feed wasn't recording his actions in the disaster. But someone was always watching. That's the way life worked. Someday, he knew, there would be a reckoning.

2

Cruise

I

"I remember looking up at the sun, closing my eyes, just letting the heat soak in." Hadley Nobile kept her eyes closed, a smile breaking across her face. "And of course I loved our autumns. The changing of the world. The warming of its colors as that orange and yellow blush painted all the trees. But you could feel it in the wind, you know? You remember how it was: the tentativeness. The warning. October showed up and the snows began to fly, and by November the skies turned that iron gray." The smile was gone. "By the time February came around, I was hanging on by my fingernails, gritting my teeth, cursing the weather gods every time I had to shovel my driveway or fight rush hour traffic on the slick roads. It's different when you're a kid, but once we grew up, spring could never come soon enough."

Gibson van Clive shifted uneasily, brushing a hand over his prematurely balding scalp. "You sure picked an odd place to do research."

Hadley looked out the porthole, smiling again. There was no bitterness in her expression, but a flame of excitement glowed in her eyes. "Make no mistake, Gibs, Europa chose me."

"I think Europa chose all of us," boomed the voice of a large man as he entered the galley. The room was not spacious, and seemed less so with his formidable presence. "Only a fool would spend months traveling out to an ice ball, and look at us: nine idiots gladly going, voluntarily racing out to the cold darkness."

"Ah, the ice man cometh," Hadley chided.

© Springer International Publishing Switzerland 2017
M. Carroll, *Europa's Lost Expedition*, Science and Fiction,
DOI 10.1007/978-3-319-43159-8_2

Gibson joined in. "Did you say cold? Could it be that our good Dr. Sigurðsson, who spends six months a year in the Icelandic winter gloom, is being a temperature wimp?"

Orri Sigurðsson's blonde eyebrows looked as though they had met at a shrubbery convention and had never quite gone their separate ways. They bobbed up his forehead conspicuously. "I got acclimated to this balmy ship." The man gestured grandly with palms up and joined the others at the single large table in the center of the room. Three small tables huddled against the wall, but the nine explorers usually left those for the handful of tourists on board. The giant smiled at Hadley. "Hey."

"Hey Orri." She jutted her chin toward Gibson. "Ignore him. I always do."

"You'll have to teach me that trick."

"You're just a science nerd," Gibson said dismissively.

Hadley glanced toward the doorway. "And speaking of which—"

Sterling Ewing-Rhys approached the table on cue. Short, muscular, quick, he was a fireplug of a man with graying temples beneath an impressive head of slicked-back hair. Someone you wanted on your side, not on the other. His name conjured up English accents and Earl Grey tea, but in fact he came from just south of the Mason-Dixon Line.

"Good evening all," he said, taking a seat next to the Icelander. "Guilty as charged. I wear the label of science nerd proudly. The Sun is shining. The birds are singing. Lovely day, isn't it?"

"If you like black skies," Hadley said. "And digital birdsong."

Dakota Barnes sauntered in, trailing Joel Snelling and Ted Taaroserro behind her. It seemed only natural that the stunning Dr. Barnes would have men trailing her in her wake; it came naturally to her. But Gibson thought she didn't notice, or at least she didn't flaunt it.

Ted had to duck to get through the door. The poor guy was always ducking, always banging his head on something. Gibson tried to imagine the man in a black frock, perhaps tending an orphanage, but it was no good. He knew too much about the pastor's research to envision him anywhere but in a high-tech lab.

"Did you save some for us?" Dakota said, looking down at the table. She stuck out her lower lip. "Oh, has there even been any yet?"

"No sign of vittles to this point," Gibson said, "but we have high hopes."

Dakota sat. "This bucket could probably use more than one steward to take care of us."

"That's all the research grants allowed for," Hadley said. "Just have to be patient. I'll bet the kitchen would let you pitch in."

Dakota shook her head. "You wouldn't volunteer my services if you knew what kind of cook I am. Any sign of the lovebirds?"

"Amanda said she and Aaron were giving dinner a miss and they'd see us at breakfast," Orri said.

As the three sat, Hadley glanced around the room. This was her rodeo, and it was time to get things organized. "Okay, gang: everybody who is going to be here is here. So I'm sure you're wondering why I called you all together like this." Courteous laughter scattered around the table. "I do have something I wanted to say when we were all in one spot, which doesn't seem to happen much, what with everybody's different shifts. Now is as good a time as any."

"We won't lose our appetites, will we?" Gibson asked. One of the tourist couples entered, saw the boisterous group, and quietly disappeared back into the hallway.

Hadley smiled wanly. "Nothing so serious. I think we've made a lot of progress as a team, and we've still got weeks to our destination. Thanks for all your hard work on that count. We've gotten to know each other, people have stopped calling me Dr. Nobile or boss—except for Ted, who calls everybody boss."

"That's right, chief," Orri put in.

"Or chief, and we've gotten to know who is best at ping pong and 3D chess—I'm never challenging Ted to a ping pong match again—and I was just thinking we should share any concerns or action items now, so we have some cruise time to get ahead of any problems."

Good, Gibson thought. She's showing her leadership qualities. He knew she was up to the task. He just wondered if she knew. He had known that crystalline brilliance of hers, from the time they were in school together to the remarkable rise of her career. She had certainly proven herself as a virtuoso in geology, but not so much in the nuances of human nature.

Ted Taaroserro cleared his throat, interrupting Gibson's thoughts. The man's voice matched his build, thin and reedy. He spoke precisely, like an Oxford scholar or a person trying to mask the fact that English was his second language. There was something earnest in his manner, something transparent, endearing. "I just got the update from those engineers up on Taliesin. Our submarine is on the crane and ready to transport to the south base as soon as my team says it's good to go."

Hadley pursed her lips in thought. "You're not selling ice cubes to this Eskimo, are you?"

"No, really boss. All is ready, far as I know. I've checked magnetosphere patterns—"

"For Jupiter or Europa?" Gibson asked.

"Both, since they are different. Jupiter's come from its core, while Europa's are induced by its oceans as the moon moves through Jupiter's fields. And both are looking quite benign for our purposes at this point—certainly manageable on the leading hemisphere. They are, of course, in flux. The south dome is secured and pressurized at Sidon Flexus, just waiting for us to arrive. And that distance from Taliesin to our Sidon Flexus dome—all 850 klicks of it?—they're saying it only took their ground party eight hours to get there. Pretty flat terrain if we follow their markers."

She looked at him skeptically. "I'm familiar with your unbridled optimism."

His grin was his only response, bright teeth showing from his rich mahogany face.

Hadley nodded, apparently convinced. "Good, good. And what is it going to take once we're on site?"

Ted shrugged. His shoulder bones protruded against the fabric of his long-sleeve T-shirt. "If everything's working—which it usually isn't on any complex machine—it should only take us a day to actually prep the thing. But we'll of course need to have Orri and Gibson study the latest charts from the robotics before we go down. See how the ice floes have morphed, where the subsurface channels go, and so forth. These chaos regions can be tricky. Orri and Gibs get final say."

"I know you're anxious to get in the driver's seat of that thing," Hadley said. "No need to rush into anything."

"Especially after the last one," Sterling said.

An uneasy quiet settled over the group.

"No worries," Ted said, sounding a note of encouragement. "That's ancient history. Pre-war technology. We will have no reruns of past shortcomings. We've had a decade of improvements since our regrettable world conflicts. Peace now. And a prosperous science program."

"Yes, my friend," Sterling said, drilling Ted with his eyes, "but nobody knows what happened to them ... still."

Ted seemed undaunted. "Look, our redundant systems have so much redundancy that it's not even worth saying again, as it would be redundant. Lots of backups. And we know the area well from ground-penetrating radar. We're good to go, boss."

Hadley nodded. "Joel?"

Joel jerked as if he had been caught doing something illegal. "Yep, should only take a good long day. As soon as we're on site, I'll power up various systems and check out the claw, things like that."

"Sounds scary," Dakota said. "The claw?"

Joel smiled self-consciously. "That's what we call the sensing arm. All those instruments are just what you're going to need."

Dakota sparkled. "Claw discovers cetaceans on Europa. I like the sound of that."

"Spoken like a true marine biologist," Ted said. "But Amanda won't let you forget the little guys. Microbes may be the spice of life on Europa."

"Hey, if I get big critters, she gets her little ones, too. Let's not be selfish."

Joel continued in a humorless monologue. "The report I just received from Taliesin said the shaft has been dug all the way down to the level of several Schmidt chambers, so we can use those underground lakes to make our way to the main ocean, as long as the ice hasn't shifted substantially at that level. I'll check out the structural integrity of the borehole at south base, and then we're good to go from my end."

"And get this," Ted said. "Someone was actually listening to us: the shaft has been drilled at an angle—my suggested angle—of thirty degrees slope. We'll be able to slide that sub down slicker than lightning without putting too much stress on the winch."

"I concur with Joel's timeline," Sterling's southern gentleman's lilt bleeding through. "I can have human/robotic interfaces up and running in no time. Before, during and after. No matter what kind of biology there is or isn't." He directed the remark at Dakota, a glint in his eye.

Hadley took in a breath and let it out, sounding satisfied. "Good. Other concerns? Now's the time, people."

"I'm concerned about starving to death," Sterling said. The food arrived just then.

Gibson took mental inventory of the group: he found Ted the most intriguing. Here was a living contradiction, a man of faith who had made a name among the most critical-thinking scientific minds. Ted's credentials were legendary, his expertise renowned in both submersible research and magnetospheric studies.

The big Icelander, Orri, seemed at times more like a teddy bear than a world expert on glacial flows, both terrestrial and on ice worlds. Orri's work on the ice flows at Pluto's Sputnik Planum had nearly earned him a Nobel for robotic research in an existing field, but his fame never outpaced his humility. He seemed happy to be among colleagues exploring nature's wonders.

Dakota may have been the most enigmatic to him, although he never had been good at understanding women, according to Hadley. She was probably right. He found Dakota's youth and energy physically attractive, but there was so much more to her. Like Ted, she fought against stereotype. In her case, her physical beauty detracted from her passion and brilliance in microbial biology. Like the rest of the world, Gibson and Hadley had first noticed her through her work on Martian microorganisms. She was the perfect fit for an expedition to the seas of Europa.

Which left Sterling Ewing-Rhys. When Gibson had first seen his bio, the term "human/robotic interfaces" seemed a bit pretentious. But then Gibson got a chance to read about the guy's work. His invention of cranial implants enabled operators to directly interact with remote devices on a level never dreamed of before. Sterling was the real deal.

They were uniquely suited, each in his or her own way, for a voyage of discovery. Every one of them saw the physical world as a shell, a mask of a greater truth beneath. Under that mundane surface, a place of tedious interplanetary travel, of taxes and electronic filing and three meals a day and the relentlessness of sunrise followed by sunset, other things waited. Exciting things to be discovered. Verities and beauties and excitement to be revealed. That's what made a good scientist. That's what drove everyone here. Everyone, perhaps, except for Aaron, but as long as he was a good physician, Gibson didn't care about his outlook.

II

Gibson sat at his desk, massaging a tablet and struggling to see its corrupted data. It was difficult, frustrating work, but Hadley had entrusted him with the old journal, and he would make her proud. So much had been lost from the first expedition, the "lost expedition." Skeletal details were all that survived of the official records. Some had been lost during "the War," as had so many things, but for the most part the lack of data was simply due to poor communications links in those days.

The War. Gibson mused at the term. Nobody called it the *Third World War* or *the last war*. The global conflagration had been so much more than what had come before that it didn't bear comparison. Instead, people simply referred to the conflict as The War.

The lost expedition had taken place nearly a decade before Europa's communication satellite constellation had become operational, back when pre-war Taliesin was just a handful of habitats and Ganymede was mission control. The tablet was, even in text and with all visual or audio enhancements missing, a real fossil.

Gibson's studies of the icy satellites made him a sort of historian, a keeper of the records of the Solar System's formation. Studying a primordial crater field or cryogenic outflow channel was akin to studying an ancient manuscript. Geologists even referred to the ghost craters on the icy moons as palimpsests, in a nod to old parchments that had been scraped off for reuse, retaining the ghostly image of what had been written before. Here was deep history. It made him feel connected, as if he was part of something bigger, a great river of time. That river flowed one way, unless you had a little dinghy to paddle upstream. That's what the oldest surfaces of the ice worlds provided—transportation back in time.

The decrepit digital diary was a ticket upstream as well. It was an old piece of hardware, pre-war technology, but Gibson felt sure the text entries within would be worth the effort.

> —*Day twelve on site*
>
> *Ice core is now 450 meters down. Peter thinks… will be able to break… ough within the week, but the radar guys, esp Donnie, aren't so sure. If… and we just… miracles do happen…*
>
> *Construction progresses on our little project to the north as well. Apparently the biology contingent doesn't have enough to do (yet). Donnie and Genevive are the ones behind it; they were first on site to set up our Science Station B. Looks to me like it's all beneath Mendelson. He seemed really angr…J… not just for fun. Claims it's good science. Will have to see to believe… it IS therapeutic, no doubt, and at this point there is a lot of value in that. Leave it to Donnie to figure something like that out. The guy is creative, I'll give him that. As for…*

The data degraded into confused characters on the monitor. Day twelve on site. The spotty records indicated that the expedition had left Taliesin on February 6. With a five- or six-day travel route, that would have made the date of the text February 12, some two weeks before the end. As the entry faded into corrupted gibberish, Gibson stared at the names. They represented real people, scientists not unlike himself. Genevieve Dupre, glaciologist. Donnie Ramirez, radar/subsurface studies engineer. Peter Kaminsky, biologist/deep core expert. Dave Culpepper, fields and particles aficionado and writer of the diary. And the three others, all world-class experts in their fields. All lost. The list reminded him of his own inventory of Hadley's expedition. He shuddered.

The door chime derailed Gibson's train of thought. He glanced at his monitor. The I.D. read "Barnes, Dakota."

"Come on in," he called, putting the antiquated tablet into sleep mode and sliding it under his pillow.

Dakota sprang through the door in that energetic way that seemed to mark everything she did. Gibson studied her, a ping pong ball bouncing into the room, crowned by a golden tiara of hair. She provided a stark contrast to Hadley, who he had spent long, intense hours with over the past week. Dakota was more than carefree; she was uncomplicated.

"Hiya Gibson, you 'cool guy.' What's the latest on the ice moon front?"

"No mucho. You?"

She tapped the wall next to the doorway. "Just living the dream on this luxury liner. I was trying to work out, do one of my katas, but my little quarters felt too cramped."

"I didn't know you practiced martial arts," Gibson marveled. He surreptitiously turned his attention to her muscle shirt and sweat pants. People tended to dress less and less formally during these long trips, but now her ensemble made more sense. Even beneath her soft leggings, he could see the outlines of chiseled muscle.

"Never came up. I like to do it in private. Ever been on one of the Princess Cruisers, Earth-to-Mars?" She shook her head with a faint smile. "Those things, I tell you. Amazing. Fully equipped gyms and buffets to die for. But we do get our own quarters here, which is more than I can say for the last barge I was on. Earth to Mars and back. Cheaper than Princess, but brutal."

"Are you a First-Class traveler?"

"Coach is fine as long as you have room to turn around in your bedroom without flushing the toilet. Those barges are cramped for a flea."

Seeing no other place to sit, she plopped down onto the bed. Ping pong. She brushed the bedspread with an outstretched hand, scooting back to lean against the wall "They were stingy with your furniture, weren't they?"

"I stuck my other chair over by the porthole."

"Oh, yeah. By the way, I like your bald head. It's honest."

He hid his surprise—and amusement—behind conversation. "You know what I like? I like the fact that when it comes to Dakota Barnes, everybody gets what they see. You don't pull any punches."

She shrugged. "Life is short. Don't change the subject. I'm just curious why you didn't go the genetic supplant route."

"My two uncles did. Bald as bowling balls, both of them. But my sister told me the procedure made them look like actors. I wasn't sure exactly what she meant, but her tone told me it wasn't something I wanted."

"Good decision. Hey, what is it with Joel? The guy is wound up tighter than an ion drive. He grumps about everything."

Her bounce to another subject would have been startling, had it been anyone else. Gibson weighed the moment. How much history did Dakota know? She was young, and certainly was an inexperienced explorer. "Joel has been through a lot."

She seemed to not hear him. "When it comes to worrying, he's an Olympian. It's like he trains for it, practices long hours, honing his skills until they become second nature. He's remarkable."

"He does seem to fixate on things a bit."

She held up her hand. "That sounded petty of me. Sorry. My father was a worrier, and I saw how it distracted him from the good things. I guess I don't want Joel to miss out, you know what I mean?"

Gibson held her gaze for a little too long. He glanced away. "I do. Yes."

"You don't think he's getting the shakes, do you?"

"Doubtful. He knows the symptoms of deep space narcosis, and we're all watching each other. But he's been out here before. DSN probably would have shown up in his earlier travels, and he was stable."

"Physically? Emotionally?"

"Both."

Dakota nodded, keeping silent for a moment. "So I had a vision."

He frowned at her skeptically. "A vision? Like Joan of Arc?"

"Nothing so religious," she replied demurely. "That would be Ted's department. I should say I *have* a vision, for this expedition. It has to do with in situ life on Europa."

"Imagine that, coming from a marine biologist."

She looked up at the ceiling. "Please! I am really excited about this stuff, this place, the prospects."

Gibson surmised that her youthful enthusiasm was taking her too far ahead of things. It was clearly early in the game, with lots to learn. He leaned toward her, hands on his knees. "Dakota, it is an exciting prospect, exobiology and all. But right here, in this room, just you and me, can you tell me you really expect to find some sort of advanced life forms out there?"

"Science does not deal in—"

"Expectations," he finished for her. "I know, I know. But what are the chances? You're here for the duration, got your spot on the team, so be honest. Fifty-fifty? Fat chance? Nil?" His tone teased.

"Doesn't matter, really. We'll do our best to find out, won't we? In my humble opinion, if we don't find it on Europa, we ain't finding it anywhere in the outer system. The stakes are high."

"Steaks sound good right about now. A little garlic, a few diced onions..."

"Wrong kind of fare." An intensity charged her posture, an excited tension. "See, everything we know about agriculture, everything we know about ocean floors and organic sedimentation, about carbon cycles and about medicine, it's all based on one biome, one family of living forms, one set of biological rules. It's as if we've had vanilla ice cream all along without even knowing we could compare it to chocolate or strawberry or Rocky Road."

"Hey, I like vanilla ice cream."

"You and I have so much in common." Her eyes sparkled as she said it.

"Besides, what about those tiny Martian beasts you've been studying?"

Dakota frowned. "My money's on those having a terrestrial origin. RNA structure similar to ancient Haloarchaea. Living in a salty biome. No organized nucleus, similar enzymes involved in translation and transcription—sorry, does this mean anything to you?"

"I'm keeping up," Gibson said, "just barely."

Her expression lapsed back into thoughtfulness. "So if all that is true, we have one book of life. But what if there's a whole library out there?"

"And Europa might be our first new book, just waiting to be read?"

"It would crack the entire field open. Life independent of solar energy, purely chemosynthetic. In a place besides Earth's ocean floors. Imagine how the prospects for life in the universe would change. We would go from thinking about habitable zones around stars to thinking about vast arrays of icy worlds well outside of those zones. Sun-like stars? No need! As long as there's an energy source, like radiation or volcanism or certain chemistries, maybe life thrives. The possibility of biology in the universe would skyrocket!"

"And if not?"

She shrugged. "If not, it doesn't prove anything. There still might be active biomes inside Enceladus or some of the Uranian satellites. Maybe Triton out at Neptune. Who knows? But the affirmation of just one more example would blow the lid off."

Gibson felt himself grinning. Her verve was contagious. "Yes, pretty exciting stuff."

Dakota pulled herself to the edge of the bed. "More exciting than those great ice cores?" She poked his nose impishly. He tried not to look down the front of her sweater, but she did have very nice curves. "More exciting than those beautiful glacial flows?" She inclined her head toward Gibson's, her eyes staring into his. He could feel her breath, smell cinnamon and something else. Jasmine. "I mean, it doesn't get much better than watching a river of ice advance for a few millennia, now does it?"

She was beautiful, athletic, youthful. He wondered if her skin tasted like jasmine.

His wife had always loved jasmine.

And then the darkness returned.

Gibson blinked. He sat up, tilting away from her. "Can I tell you a story?"

She sat up, too, surprised. She brushed her hair back and assumed a patient lotus pose on the bed. "Tell away."

"Did you know that I was married?"

"Married?" she barked. "They've got your file as single."

"Now, yes. I was married before, I mean. To a wonderful woman. My soul mate, really. She died in a suborbital last year. Flying the route from Syrtis Major to Lowell City."

"I heard about that crash. I'm so sorry."

"No need. I'm dealing. If you have to check out, I guess Mars is a beautiful place to do it. She certainly loved it. Geologist. Like me." He took in a deep breath. "But I just haven't been the same since. It's all pretty raw, and I guess I'm not really ready for, you know, anything Shakespearian." He couldn't look at her.

"Are we talking tragedy or comedy?" She said it gently.

He looked up. "I was talking romance, but I suppose either of the others fit."

A blush made its way across Dakota's chest. "I'm really sorry, Gibson. I didn't know. Here I am, pushing myself on—this is not my modus operandi."

"Now look, there's no need to explain. These long trips get quite lonely, and we've got to be sharp in the field so nothing goes wrong. Stay frosty. Maybe after it's all over ..."

"Sure, yes." She said it in a calculated manner. As she stood, she patted him on the knee. "I'll just head back to my room. You know where it is."

Gibson stood up in front of her, blocking her way. He kissed her, gently, on the forehead. "Believe it or not, you made my day," he said.

"Well, that's something, at least," she cooed.

She passed through the door, her muscles as taut as a ballerina's. "See you at breakfast," she said as it closed.

"End of a perfectly lovely evening," he muttered.

III

"It won't do, you know."

Hadley sat at her desk in her darkened quarters, and knew who it was without looking behind her. He always came to her in the darkest hour, the lonely time in that blackness between midnight and the hopeful dawn.

"You don't know that," she said into the darkness. From the shadows, he emerged. He wore what the early adventurers used to call an Eskimo parka: Inuit design, fur-lined, with collars and sleeves of fluff spewing from wolfskin leather.

"My dear, my dear. Your presence won't make any difference. History has its own inertia, a relentless march. There is blood in these things. Blood and loss and heartache."

He pulled his hood back to reveal a military crush cap, a general's eagle enwreathed within golden laurel upon the front.

His oversized mittens held the pattern of leopard skin, the spots a deep crimson. It was not the color of blood, but something else, a thing she couldn't quite remember. If her artist brother were here, he could tell her. He knew all those reds: alizarin crimson, napthol, cadmium red, quinacridone. All she knew was that his hands looked as if he had shoved them into a Thanksgiving cranberry sauce.

"You can't blame a girl for trying."

"Nobody is blaming you for anything. Are you blaming yourself, my belladonna? Ahead of time, for something that may not happen? Or for something past?"

She shook her head.

The phantom crossed the room to the window. "It would simplify things, of course, if you let it go. Many have had to. There is freedom in that."

"I can't. You know that. I have to know."

"At what expense? At the expense of your friendship with Gibson? At the expense of your life?"

He began to fade, holding a hand up. Was he bidding her goodbye, or warning her? She shook her head again. Something brushed against her lashes. Her pillow.

She sat up in her bed, rubbed her face, and stepped to the bathroom. It was a short journey on this ship, in these quarters, but she wasn't complaining. This was a luxury liner compared to some of the tubs she had traveled on, especially out this far.

Was it possible for disaster to run in families? Could it be that bad luck, or fate, or evil influences were passed down, generation to generation, to attract catastrophe like a lightning rod? Perhaps some of Scott's descendants perished in icy graves. Maybe the *Titanic*'s captain passed the sinking-ship-gene down to his grandchildren's children.

Hadley splashed cold water on her face, decided it was a mistake, and turned the faucet to warm. That was better, but the chill wouldn't go away.

IV

At first, it looked as though the Grants would be no-shows for breakfast, but the married couple of the team finally made their appearance. Amanda and Aaron Grant entered the galley, with Aaron perennially looking like a sullen child, not particularly happy to be there or anywhere else. Amanda wore the same smile that she plastered on her face for any group occasion. When they weren't around, Hadley referred to them as the "odd couple." Gibson liked that. Physician and microbiologist. Frick and Frack. He knew Hadley didn't share the thought with anyone but him, her confidante. Speaking of Frick and Frack. He smiled to himself.

The couple sat at the table, in adjacent chairs, but it seemed as if Aaron carved out as much personal space between himself and everyone else as was possible in the confined galley. "Morning," Aaron said quietly, his voice an insecure tenor. His tentative tone seemed to match his pale skin, which couldn't make up its mind between being a bleached Caucasian or simply translucent. His sable hair, undoubtedly dyed or genetically supplanted, merely emphasized his waxen complexion.

Of all the researchers on the team, Aaron and Amanda puzzled Gibson most. Despite the group activities and the science strategy meetings and the team building sessions, he just couldn't get a bead on them. What made them tick? Why was Aaron so odd? It was only natural that Amanda would shelter him if she felt protective as a spouse, but Gibson felt the need to be closer to them, the way he felt with his other colleagues. He would have to do something about that. Perhaps tomorrow.

"Hello, team!" Amanda said energetically, her full head of salt-and-pepper hair bobbing in the 1/4 g of the ship's acceleration. That rich silver mane seemed to rest uncomfortably atop the careworn face. Amanda had clearly had some work done on various wrinkles and sags, but her age was showing anyway. Gibson thought she would probably be more beautiful if she had left her face to age naturally. Like cheese. He never understood that side of some women, the desire to stay forever young. Then again, guys played out that desire in their own way, doing stupid things beyond their age bracket. What women did to their own aesthetics, he supposed men did to their own athletics. Either way, it was usually ill advised. Gibson came from a long line of prematurely bald men (and women, if you counted his great grandmother Edna). He was definitely headed that direction. Bring it on. Besides, people like Dakota didn't seem to care.

"But it's the surface we need to worry about first," Orri was saying in his basso profundo. "And I guess if we're voicing concerns, that one's mine. Chaos regions are notorious for shifting surfaces, breakthrough, crevasses opening up, slip faulting and the like. And as we're headed down to the site during the most unstable part of the orbit…"

"By design, obviously," Aaron cut in. He seldom contributed to the meal-time conversation, withdrawing into monosyllabic responses to the surrounding chatter. It was a shock to hear his voice. But he was right. Europa's geysers could only be observed at one spot in its orbit, and even then the phenomenon was not a given. Both Io and Ganymede had to be in certain spots in their own orbits to tug on Europa's crust in just the right way, and they needed just the right help from Callisto on the outside. Further, the triple arrangement had to occur at a precise spot in Europa's own orbit, which was only slightly out of round and inclined to Jupiter's equator. That spot coincided with Europa's greatest distance from the planet. All of the factors had to come together in a rare perfect storm. The expedition would only get one shot at this, or the researchers would have to wait for years. By then, the scientists would be long gone.

"Leave it to our physician to put a finger on the celestial mechanics issues," Hadley said with admiration. Aaron offered a weak twitch at the corner of his mouth in return.

"That's the tension, isn't it?" Gibson put in. "We need a place thin enough to drill through, but stable enough to set up operations. We did some work on Enceladus a couple years ago, and had no trouble at all with surface stability."

Orri tapped his finger on the table. "Yeah, but as you know, Enceladus ain't Europa. Whole different ball game. We're talking mostly southern sea versus Europa's deep ocean. And thick crust vs. thin, ongoing geysers versus occasional and rare."

"But ice is ice," Gibson insisted. "Even with the temperature differences between Jupiter and Saturn, the ice behaves essentially the same way. And the ice at Enceladus's south pole is every bit as unstable, at certain times in the orbit, as that of any chaos region."

"I wish I shared your optimism," Orri grumbled. "On Enceladus the fractures can penetrate much more deeply than on Europa, as gravity tends to hold fractures closed at shallower depth on Europa."

"True," Gibson said. He began to comment, but Orri continued. "And Europa is more like a glacier than Enceladus is, generally speaking. It's a surface of ice rivers and subducting blocks; drill into that stuff and you've got issues. You know the problems we've had in the past." He looked down at the table. Gibson noticed that everyone was avoiding eye contact with Joel. Joel was studying the table, too, and Gibson spotted a sheen of sweat on the man's forehead, a red blush running up the sides of his neck.

Hadley chimed in, "All right, I suppose the take-away is to go easy until our glaciologist-ice-moon-guys determine that things are stable enough to proceed. We get into the hot zone, observe the geyser activity, establish our remote science stations, and get back to the dome to get into the ocean quick, yes?" She looked at Orri.

"Tough call," he said. "Might be best to do a dive or two before we do the geyser leg of the expedition. Just to make sure we get everything. But I'd rather not decide that on my own."

"You'll have help from other experts," Gibson said.

"Yes," Ted added, "They say there is wisdom in many counselors."

Dakota leaned over and put a hand on Orri's arm. "We're in this together."

"Yes," Amanda said. "Together for strength."

"Together for power," Aaron added reflexively. He shot a glance at his wife. Amanda seemed to pause. What did Gibson see there? It had been subtle, just the slightest hint of acknowledgement, the twitch of the mouth, a forced blink. Was that panic in her eyes? He certainly saw confusion on several faces. But before he could ask about the reference, Dakota interjected, "I really didn't expect the dining experience to be this tasty. I expected something a bit more…"

"Military?" Ted asked. "I was hoping to never again have anything of the sort."

"Were you in The War?" Dakota asked.

"We all were, essentially." Everyone at table muttered agreement.

"If I was on the front lines, I'd be happier fighting next to a priest." Dakota quipped.

"Pastor," Ted corrected. "To me, it was a war unlike any other. The blurred borders of The War obliterated nationalism, wiped away conventional patriotism. I suppose wars of ideas, of worldviews, tend to do that. But yes, I fought in the Antarctic Arena."

"Whose side?" Aaron asked.

Ted blanched.

"That was tacky," Gibson scolded lightly. "It doesn't matter. Those days are over. We've all come out of it and there are no sides anymore."

"Here, here," Dakota called out. Amanda and Aaron exchanged glances again. Ted noticed.

"No harm done, Aaron. Okay?" Ted asked, looking at Amanda.

Amanda paused. "It's just that we both lost family in The War." Aaron shot her a warning look. She winced.

"In the camps?" Orri asked. "I did. A brother and two cousins. My brother died at the hands of the Seven Sisters. I tell you, Hitler's prisons had nothing on them."

"A whole factor larger, I suppose," Hadley said, swirling the cream in her coffee cup.

"Or the Third Islamic State," Ted said. "But Hitler had it over all of them for organization. The Eastern Alliance just took it to another level, with the help of the sisters."

"Septuplets?" Dakota asked.

"Hardly," Ted said. "They came from all over the place. Only two were related by blood."

"The twins," Hadley mumbled.

Gibson huffed. "Yeah, but they were like sisters in their *vision*. All played it out in their laboratories, in the camps. Birds of a feather."

"A dark vision it was," Ted said.

Orri seemed deep in thought. "Still, sometimes I feel I got off easy."

"Ah, yes, the camps," Aaron said as if just now realizing it. A strange look played across his face. It was not an expression of distaste, exactly, or horror. It was an inscrutable look Gibson couldn't quite interpret.

"The camps," Joel said quietly. "Unforgiveable."

Ted shrugged. "Oh, many people did nightmarish things during The War. But there is enough forgiveness in God even for them." He was looking directly at Amanda. "Forgiveness for anyone, no matter what."

Gibson watched Amanda. Only Hadley seemed to notice what he saw. The spindly woman had a deadly energy about her, like a coiled spring. Amanda stared at Ted. A fire kindled in her eyes that he had never seen before, a burning malevolence. If she had been a snake, he guessed she would be hissing.

"Believe it or not, I was a medic in The War," Sterling Ewing-Rhys said, breaking the awkward pause. "On account of my experience with cranial implants, I suppose they thought I was qualified." He tacked an extra, drawn out syllable onto the last word. "I have seen enough and thought about too much. I'd rather leave all that behind."

"And I'm sure we are all moving on from those dark days," Dakota said, lifting her juice glass. "Here's to the end of war and the rebirth of research, to new horizons and exciting discoveries, no matter where we're from!"

The orange juice glasses and teacups clinked, the coffee warmed and soothed. But most of the diners didn't stay for a second cup.

In the corridor, alone after the group had dispersed, Gibson asked Hadley, "Why didn't you tell them? Why did you leave that little tidbit out?"

"They'd had enough of my family history."

"Ancient history, yes. But you left out the more recent stuff."

"Just wasn't the right time. If it comes up again, I'll say something."

Gibson shook his head and headed for his room.

V

The dinner's conversation loomed from the walls of Hadley's empty room, urging her to remember. The camps. The battlefronts. Memories of The War plagued everyone differently. Each person had a story to tell, a tale of loss or pain or triumph or heroism. Hers was a story of loss. She had always reveled in the relationship her mother and father had. They fought like any other couple, but they fought fair and they always made up. The parents she knew in the very private family times were the same ones the public saw. Her parents were well-loved, and they loved each other. They loved Hadley and her brother. And when The War came, her mother and father became the solid ground in the maelstrom, despite the fact that Hadley was already on her own, making her way through the scientific community.

Hadley stared at the screen, at the old photo of the polar explorers. Umberto Nobile—another link in the Nobile chain—held his dog, Taita, as he posed before the gondola of a huge dirigible. In the sienna-tinted black and white image, it was hard to tell if the sky was clear or overcast, but Nobile's companions looked cold. She wondered about them, wondered if they felt any differently about their expedition at the outset than she and her companions now felt. Were they excited? Did they hold high hopes of discovery and even adventure? Were they passionate, like Orri, who couldn't wait to get out on an alien glacier? Or like Dakota, who was dying to get her hands on some real Europan water? Did they have reservations about safety? Fear of failure, scientific or otherwise?

She shook her head. It wasn't Nobile's expedition that haunted her. It was a much more recent one. What was she to do with the memory of Donnie Ramirez? Of David and Genevive...

The door chimed. She tapped a key and the screen identified her visitor: Ted Taaroserro. "Yeah?"

The outside entry opened and Ted stuck his head through. "May I?"

"Entrée, Dr. Taaroserro," she said, misusing the word playfully. She gestured to the chair by her desk. "Make yourself comfy. So what brings you all the way to this side of the ship?"

"I wanted, first, to ask your forgiveness for my actions at breakfast. I have already spoken to the affected parties."

For the first time on the trip, Ted Taaroserro sounded like a man of the cloth. When he had first signed on, Hadley had been nervous about including a religious person on the team, but Gibson and others knew of Ted's track record, his expertise in energy fields and particles, and his fine history of sea-ice submarine research. His addition to the group turned out to be a feather in Hadley's cap. She turned her attention to him. "Affected parties? What are you talking about, Ted?"

"I was unkind to Amanda and Aaron, I think."

"I suppose you snapped a bit, but I didn't notice any unkindness."

"Perhaps it's that I know what was in my heart at the time. This kind of long-term travel sets me on edge. But at any rate, I also wanted to tell you how much I appreciated dinner last night. Just the way you aired things out. Gave us all a chance to sort of put all the cards on the table, since we're all playing the same game."

There was something he wasn't saying. She looked at him long and hard. When only silence came, she said, "Yeah, but are we playing Go Fish?"

Ted paused, furrowing his brow. "More like Poker, I think. And I was bluffing a bit."

"You always were a good salesman." She gave him time, waited.

Ted laced his long fingers, placing his elbows on the table. He looked down, as if embarrassed. "While I have every confidence in our submersible technology, it's the landscape that I have my doubts about. It is a rare thing that I find myself agreeing with Aaron." He locked eyes with her. "Look, these geysers at Sidon Linea turn on for a few days, sometimes a couple weeks, right? That's it. You guys do your 'Old Faithful' watch and then we leap into action because the ice is at its most pliable, theoretically."

"Theoretically," Hadley encouraged him.

"And the ice is also at its most unstable, yes?"

"Right, as Aaron said. But you've read the research, seen the reports by Orri and Gibs, right? After those geysers shut off, things settle down quickly."

He was holding his hand up defensively. "Sure, sure. I'm sure everything's going to be just hunky dory." He shook his head. "It's just a little scary, isn't it? Being six or seven hundred kilometers from the aid of Taliesin base and all."

"Closer to nine."

"Thanks a lot. Those few hundred extra klicks could make a lot of difference if we were walking."

"What, you think you're going to need to set off across the frozen tracts like Amundsen and his dogsleds? If anything goes glaringly wrong, they can fly us out on a long-range jumper. The base has a small flotilla of them."

"That's encouraging." The flat tone in his voice betrayed his true lack of conviction.

"I find it so."

"Right. But aren't the majority of the jumpers one-person affairs? That would be a difficult ride back. And we add another one-fifty on the road to get to the geyser observation site. A thousand kilometers of icy vacuum is enough to give one pause."

"Getting cold feet?"

"It's going to be a lot more than our feet that get cold if…" He let the thought die in the air.

Hadley leaned forward, a sly smile gracing her lips. "And if it wasn't a long way, if there were no risk, it probably wouldn't be worth doing. That's what we do. We explore." She felt her pulse race as her first love came to the fore, the joy of pure knowledge so strong that it lay in her mouth like a taste. "We fight against those boundaries; we cross the lines, run off past those places where our maps say 'Thar be dragons', right?"

Ted looked troubled. He fidgeted like a cornered mouse. "All we can do is give ourselves as many Plan B's as we can, and trust that the Creator will carry us through." He wrapped one hand in the other. "You would think I could get used to this, exploring around dangerous places. I've been on a lot of expeditions. I did some polar exploring on Mars, you know."

"I thought you were an Earth guy, top to bottom."

"Good Masai stock. That's me, boss. It was after The War. I went to Mars to get away from things. You know how it was."

They both sat in silence for a moment, sharing the loss, the ache, the sadness.

Hadley said, "Don't worry about the jitters. It happens to everyone on deep space missions. There's some powerful thing that happens inside you when you're this far from home, something that comes from deep down."

"From that reptilian part of our brains, I suppose. The built-in nature that we have to combat daily."

"Or at least come to terms with," Hadley added.

"I am being irrational, I know. But it just sneaks up on you. Fear of the unknown and all that. Still, God is in it. That makes it all worthwhile."

"In what way is God in this? I'm just curious." Hadley's tone lacked any antagonism.

Ted stood and stepped to the little portal. The brutal light of an icy day shimmered upon his ebony face. "We're going out into the great unknown, into His creation. That's a beautiful thing. Whatever we find out there will have His fingerprints all over it. To know a creation is to know its creator, don't you think?"

"I suppose so." She furrowed her brow. "You find beauty in that creation?"

"In all of it. In you, too. People, far more than the inanimate. You see, you were created with intention. You were thought about, born in God's thought and then created. His children are the grandest and most precious of his creations. And that, Hadley, is who you are. I am glad we have redundant systems and grand plans, but those plans are not where I find my comfort."

After a moment, she said, "So, is God on our side in this?"

Ted turned back to her, smiling softly. "God 'does not wish anyone to perish.' Of course, that quotation was speaking spiritually. Everybody likes to think that God is on his or her side. I'm sure that was the opinion of all parties in the last war."

"Sure. But how could He be?"

"Are we speaking of world politics now, or exploration?"

"Either. Both. I don't know."

"All I know is that the Creator has a plan, and it is good. Sometimes we can't see it with all of life's mess, but I am convinced it is true."

"You get paid to say stuff like that."

A faint smile played across his face. "I speak from life experience."

Hadley locked eyes with him. "Ted, just what was that all about? That little speech about forgiveness? I thought you were being generous, but some of our group reacted rather strongly to it."

Ted blinked, but didn't look away. "Some things are better left unsaid. It is … " he searched for the right phrasing, "safer for everyone."

Hadley frowned, taking in a breath. "Well, I've had my share of wondering about this mission myself."

"And what's bothering our fearless leader, hm?"

"All the normal little things, logistics and timetables and all that."

"And something a little deeper?" Ted's gaze was clear, alert.

"You see right through me. We've got multiple assignments, don't we? Geyser activity, then oceanic exploration. But it's a bit like visiting a graveyard. Don't you think?"

"Yes, I suppose it is. I imagine you will be keeping your eyes open for whatever might be left, something the recovery teams might have missed." Ted's tone became more urgent. "But our mission is about life, not death. We'll be making our own history, not focusing on someone else's. You've put together an able team."

"And as you pointed out at dinner, with a lot better technology than they had before The War."

"Indeed." Ted's eyes shined. "And we'll have each other's backs. You're an amazing woman, Hadley Nobile. Thanks for the chat."

She stopped him with a penetrating look. "Ted, I've read your professional goals, your expedition milestones, obviously. But just what do you want to get out of this? Personally?"

He fell silent. Hadley read a host of expressions settling on his face: confusion, doubt, skepticism, even wonder. "Sometimes, to see the signature of the Creator is enough. It is open for everyone to read. You simply have to learn to recognize His handwriting."

He stood, bumped his head on the ceiling, and then bumped it again as he ducked through the door.

"We need to get you a football helmet," Hadley said. She couldn't be sure who was trying to encourage whom. As soon as the door sealed, she wilted against the wall. She buried her face in her hands. Her entire body shook. She was sweating. She had put voice to the real issue, to the subject she had been dwelling upon in her own dark corner. The lost expedition, just before The War, had plenty of Plan B's. The team members were, from all reports, seasoned explorers and careful researchers. Just like so many who had come before. Just like those on Umberto Nobile's voyage, or Dave Culpepper's. But Europa was an unforgiving place. With the distances and complexities involved, this new expedition was not going to be "just a little scary." It was going to be terrifying. Maybe with Ted around, the mysterious Almighty would be on their side after all. But the scientist in her preferred to bank on Plan B's and orchestrated alternatives than on forces she couldn't see.

VI

Amanda ushered Hadley into a surprisingly stark room. Unlike the others she had visited, this double was barren of personal items. No photos hung from the walls or graced little digital albums on desks. She saw no evidence of past voyages or homey touches in the corners. On extended voyages like this one, even the most distinguished of researchers usually brought small pieces of home, but not the Grants. Amanda and Aaron were all business. It looked as if they had no past.

The double room had a sitting area with a table, the benefits of a couple's accommodation. She would have to remember that the next time she went on a long cruise. The extra legroom almost made the lack of privacy worth it.

"Have a seat, Dr. Nobile. Aaron will be right out." Amanda sounded like a receptionist in a medical office. Neither of the Grants had gotten comfortable with referring to Hadley on a first-name basis. Amanda gestured to the chairs at the table.

Hadley sat.

Amanda disarmed Hadley with a conspiratorial smirk. "All right, girl. Spill it. How are you doing?"

"Me?" Hadley was surprised by Amanda's informal tone, but she didn't mind it.

Amanda looked around the room theatrically. "You seem to be the only one here that's leading an exotic expedition to the south pole of Europa in search of volcanoes and sea serpents. Brave girl."

"Not so sure about the sea serpents," Hadley said. "But yeah, it's a bit daunting. And exciting to have the opportunity."

"You're doing just fine. There's a whole lot to keep track of. I never was gifted at the administrative side of things."

Amanda Grant was more relaxed than Hadley had ever seen her. Her little mission to break down some barriers seemed to be going well. Amanda sparkled with energy, as if she was sharing an inside joke with Hadley. In fact, she reminded Hadley of the women in her own family, especially her grandmother. Gramma Nobile always had a way of cheering people, of making you feel like you were special and that you shared a special secret. Life was good with Gramma, and she made it so with everyone around her.

"I'm afraid I've had to be administrative before. I was in charge of comms back on Earth for a deep space expedition or two, and I even did a round on Ganymede."

"You strike me as a woman who knows her mind," Amanda said.

"That I am."

Amanda glanced from side to side, then faced her again. "We've been comparing notes with the crew, Aaron and I have. Everybody else had to endure a painful application process. It seems that we're the only members that you came after."

"You're both good at what you do, and you come as a package deal." Amanda seemed unsatisfied, but she let it go.

Aaron came in and settled next to the two women. "Afternoon, Dr. No—" He stopped himself. "I did it again, didn't I? I am used to the decorum of rank on missions. It's important to always know who is in charge, don't you think?"

Hadley felt as though she was in a Petri dish, with the good doctor studying her, just waiting for her to morph into some mutation. "Yes, to a point. But we're in such close quarters that some familiarity is called for, I think. It helps reduce the tension."

"Tea?" Amanda offered. She played mother, serving up Lady Gray to everyone.

Hadley took a sip, put the cup down, and said, "We went through all that formal background stuff when I selected you for my team, but I thought we could have a chat about life in general. It's not so easy in a large group."

The two were silent, Aaron looking owl-like. Amanda broke the tautness.

"Yes, we thought that was a nice gesture. Didn't we, Aaron?"

"Oh, yes absolutely, dear." He scrutinized the cream dispenser.

"So how did you two meet?"

"Funny story," Amanda said, straightening in her chair. "We met at a symposium in Dubai. Microbiology and effects on large populations. I heard Aaron's presentation. I went up afterwards, and he told me he had seen mine as well. I was smitten." Her tone was practiced, as if she had told the story a hundred times. Or as if she had been rehearsing it in front of the mirror.

"Smitten?" Aaron scoffed. "Is that what that was? I thought you were just coming down with a rhinovirus." They both smiled.

"Love at first site?" Hadley asked.

"Something like." Aaron ran his tongue across his upper lip.

"Was medicine always your passion?"

He glanced up at Hadley. Their eyes met, and his narrowed. "I have always been fascinated by the human body's response to external influences. Its immunological defenses, its reaction to disease and negative physical stimulus—" An energy invaded his little speech, a fervor that Hadley had never seen before in the man. He was practically salivating all over the table.

"And I've always loved microbes," Amanda interjected, "so we were a match made in heaven. Of course, I came late to the academic game. Started out in a small lab doing grunt work."

"Where was that?" Hadley asked over the edge of her cup.

Amanda paused.

"Philadelphia," Aaron said. "But I rescued her from the City of Brotherly Love."

"Yes, scooped me up and whisked me off to Europe. We made our home in Bruges then."

"I hear it's beautiful," Hadley said.

"It is." Amanda looked out the window, her voice taking on a wistful tone. "I miss the slow pace."

"Me, too." Aaron said mechanically. He seemed to stir. "Actually, I miss having my own lab. Freedom to experiment."

Amanda broke in abruptly. "What about you, Dr. Nobile?"

"What can I say? I'm a sucker for exploding mountains and rivers of molten rock."

"But surely there is more. What is it you want from this trip? This experience?"

She stared into her cup. After a long moment, she looked up. "Freedom."

Hadley shared about her education, her dizzying rise to fame as a volcanologist and cryovolcanologist, and her recent research on Venusian canyon structures. Her lectures had always been popular, but she was surprised to learn that the Grants had watched several.

Twenty minutes later, Hadley excused herself, tiring of Aaron's stilted, monosyllabic conversation. It offered all the fun of a tooth extraction. And she realized that at the end of the day, the couple had shared little, and had gained a fairly intimate history of her. They were definitely a couple, working as a team. She thought she should feel guilty about monopolizing the conversation, but she had tried. The only emotion she felt was a sense of unease.

Freedom, she had told Amanda Grant. It was funny how people could build their own prisons, brick by brick, out of ephemeral things like family history, ghostly pasts, failure or even the abstract fear of failure. Gossamer bricks, remarkably robust.

3

Arrival

I

They told Hadley that Taliesin Chaos was in one of those rare "sweet spots" on Europa. Hugging the edge of the great Sarpedon Linea band, it spread across a ridged plain of ice 900 kilometers south of the equator, a muddled landscape sheltered from the full onslaught of Jupiter's deadly magnetosphere. Taliesin was on the leading hemisphere, facing slightly downwind of the direction from which Jupiter's radiation fields overtook Europa. Just around the horizon, on the other side of the ice moon, the king of worlds bathed the surface in radiation potent enough to kill a human in an old-style suit within six minutes. But on the opposite side, the radiation fell to manageable levels.

Like her favorite pizza, Taliesin had another thing going for it: thin crust. The adjacent chaotic region had collapsed under forces beneath it, fracturing the ice crust into plates and rafts that bobbed in the slush. The rearing shafts of ice had pirouetted and pointed jagged corners skyward before freezing in place again. Here, the crust was still rigid in Europa's searing cold but thin enough to get at the ocean—or at least some connected lakes—below. If one had the patience and the technology. The outpost had the technology in the form of a great drilling rig. Some of its core samples added up to 16 kilometers in length. That tech would come in handy when the expedition pushed south, to the site of the geysers and a series of fractures that, used to advantage, would lead them to the ocean below in ways not possible even at Taliesin. Did the team have the patience? If something went wrong, if they blew it, there would not be another chance to see the ephemeral geysers for many years. Europa would sail on in its path around Jupiter, dancing away in

© Springer International Publishing Switzerland 2017
M. Carroll, *Europa's Lost Expedition*, Science and Fiction,
DOI 10.1007/978-3-319-43159-8_3

its cotillion with its fellow Galilean moons, locking its ocean beneath its crust for another decade or more.

Hadley and Gibson stood side by side, shoulder to shoulder, gazing across the frosty wilderness. The descent had been uneventful, but afforded a great view of the base at Taliesin a few kilometers to the side of where the lander had touched down. The architects had plunked Taliesin Outpost near the edge of a precipice some 300 feet above Taliesin Chaos' jumble. The cliff dove into the frozen pandemonium at a dizzying angle, ending in a talus pile of ice cubes the size of houses. Up close, the chaos took on the appearance of a stormy sea frozen in time: rubble rode fossil waves among plates and hummocks, some gentle as ripples in sand, others rearing up in tsunamis paralyzed by the cryogenic temperatures.

"Feels great to be on terra firma for a change," Hadley said, stamping her feet. "I had no idea white came in so many shades." Atop the cliff, the landscape was relatively smooth, providing the lander with an easy touchdown. The gray-brown plain undulated with the subtle rise and fall of ancient ridges, now sunken into near-oblivion. Within the gray and blinding white, speckles of crimson, azure, ultramarine, purple and dusky orange glistened in the sunlight. Boulders strewn across the landscape mingled with small impact craters, a testament to the Cormac impact crater just to the northeast that sent debris clattering onto the rock-hard surface in ages past. Best of all, floating in the eternally black sky, sat Jupiter, eight times as far across as the full Moon Hadley had seen the night she left Earth. Above Taliesin, Jupiter gazed down with its one red eye, its white oval storms and rusted cloud banners marching across its face as it clocked through phases from crescent to gibbous to full and back down again. All the while, the striped orb never moved from its perch some 20 degrees above the horizon.

They told her Taliesin was a sweet spot. But they had told her a lot of things. Like how easy the trip out was, now that they had the hybrid VASIMR engines. And how cozy the outpost had gotten with new décor and better air handlers. Before too long, they had assured her, it wouldn't seem like the last outpost at the edge of civilization, a forlorn bastion of knowledge in the darkness. It wouldn't be so bad.

"That's what they said," Hadley mumbled.

Gibson grinned. "That's what you get for listening to *them*. Whoever *them* is. You're talking to yourself again." He glanced from Hadley back out to the specular rusted white landscape. Here and there, the Sun glittered on a glacial-blue outcropping or an ice boulder. He nodded with excitement. "It's like Vatnajokull on steroids."

"Icelandic glaciers got nothin' on this place."

"Come on, now, give a little credit. Vatnajokull's the largest glacier in Iceland. In Europe, for that matter. It tries." He spoke to her, but his eyes were still plastered to the landscape.

Hadley tried not to think about the radiation bathing this winter wonderland, or about the numbing temperatures, or the fact that just beyond her faceplate stood an ever-threatening, airless void. Even for a volcanologist, it was an uninviting vista. The place was desolate. It was lethal. And now, it was home.

"Had, there they are," Gibson called.

"Where?"

"Just over the ridge, to the right. Coming around the crater rim." He held his hand to his wristpad. "Gibson to crew. You guys better get your little flat fannies out here. Rover from Taliesin is incoming."

Orri Sigurðsson lumbered around the black and gold lander like a polar bear. Hadley watched Joel Snelling appear just after him. Even his walking struck her as timid. The odd couple, Amanda and Aaron, descended the lander steps with Ted, Sterling and Dakota close behind.

A rich alto crackled in their ears. "There you are. You guys have a warped autonav or somethin? You're about two klicks west of where I expected you."

The lander's pilot thrust his head out from inside the doorway, looking less like a flyer than an apologetic kid. "Hey, I took this thing down by the seat of my pants, and they told me to take it wide. Heard you had new equipment at the base, so I gave you lots of space. But this is no two klicks away."

"Saints be praised," the mellifluous voice from the rover called out in a mock Irish accent. "If it isn't my auld friend Dirk O'Meara. But your kind concern gave me a long drive."

"Sorrrry," the pilot said, without sounding sorry at all. Over her earpiece, Hadley could almost hear the twinkle in his eye.

The rover pulled to a stop. The driver hopped out and introduced herself as Tefia Santana. After introductions all around, the crew climbed aboard the rover. Dirk was last. Hadley held out a hand. "I thought you were staying with the shuttle."

"She's smart enough on her own."

Hadley pulled him in as Tefia said, "I'm sure you're all looking forward to a little more personal space and some hot food. Dirk included, but just for a while. He can't take too much society." She sent the barb toward him in a familiar tone.

"Love you too, sweetie!"

"How many does this make?" she asked him.

"Runs? This will be my lucky seven."

"Seven runs from here to Mars and Earth," Gibson said. "You must know the way pretty well by now,"

"Can do it in my sleep."

Tefia looked back at him as she dropped the rover into gear. Nodding toward the lander, she said, "From the location of your most recent touchdown site, I think you just did."

II

On final approach to the science outpost, Hadley Nobile marveled at the scale of the site. Its construction had to be done by hand, with little in the way of robotics to help out. It had been so much easier back on the Earth's Moon. The Eurasian base at the lunar south pole, on the flanks of Malapert Mountain, had been 3D printed using the regolith. That sterile, barren lunar soil seemed so dead and dry at the time. But compared to the ice landscape of Europa, it was a smorgasbord of raw material: basalts, olivine, calcium, carbon, magnesium. For a 3D printing system erected robotically, it was a cinch to fabricate the first pressure domes, then habitats and laboratories and, eventually, the rec centers and shopping malls of an international community.

Not so on this ice ball. Everything had to be brought in or fashioned from ice blocks (reserved for the outer cold-storage facilities). Hadley admired what they had done with the place, even with a shortage of all those raw materials present on Luna.

Taliesin Outpost sprawled across 5 acres of prime Europan real estate. On its west side, a tower rose 60 meters into the black sky. From it, a cable sloped down to the ground and slithered across the landscape toward the western horizon. Hadley cocked her thumb against the window and tossed a quizzical look at Tefia.

"Power tower," the driver said. "The tether goes for 200 klicks to the west, well into the high radiation zone. As Jupiter's magnetic field sweeps across Europa, the tether gathers power and sends it to Taliesin."

"Ah, like an electric motor. Or a generator," Hadley said. "A wire moving through a magnetic field builds up an electrical current. Pretty slick."

Tefia nodded and keyed her mic. "Rover on final approach. Request airlock garage access."

Ahead, a large door slid up into the roof, exposing an enclosed area with a second rover, identical to the one in which they rode. Tefia brought the vehicle to a standstill. The door slid shut and the familiar whoosh of air surrounded them as the garage pumped up in pressure. Lights shifted from red to yellow, finally to green, signaling that pressure was stable.

The band of weary travelers stepped through the entrance. Established before The War, the facility had an unmistakable international flare. As Hadley followed close behind Tefia through the airlock and into the central causeway, she heard the cadence of Chinese, Italian, and something Scandinavian before she heard her native English tongue. The words exploded forth from a rotund, jovial fellow bounding down the corridor like a rugby player.

"Bout time, Tef. Did you say nine? We only have beds for eight. Somebody will have to sleep on the couch in the common area."

Hadley cast a glance behind her. Everyone was grinning except Joel, whose brows furrowed into a pyramid shape above his eyes.

Tefia called back to the group. "Don't trust this guy. He can't count."

Joel relaxed visibly.

"Nonsense!" the big man bellowed, holding up a massive coffee mug. "I'm up to my twosies already. Jeff Cann." He shoved a meaty paw toward Hadley, and then offered it to Gibson. The man's grip was comfortable, clearly controlled. This guy was made of muscle. "Aside from being housemaid here, I'm the submersible fanatic, so I'll want to talk to Dr. Taaroserro. I'm also an electronics guy. We all have dual roles here."

"And Tefia," Hadley said, "as head of field operations, do you have another role as well?"

"Allow me," Jeff interjected. "The good Doctor Santana is far too modest. She has multiple degrees in medicine—"

"With some boring administration thrown in," Tefia added.

"—*and* has won a boatload of awards in energetic fields and particles."

"As in magnetospheres?" Hadley asked.

"Yep," Jeff barked. "We're lucky to have her here."

Tefia cut him off with a dismissive wave. "Oh, I've found it to be plenty challenging, what with radiation levels outside, rover crashes, rolling office-chair races resulting in fractured femurs, you name it. See you all soon."

Tefia excused herself as Cann led Hadley's team to their quarters, smallish private rooms that seemed spacious after the cramped quarters of the Earth transport. Each room split off from the central causeway, a tunnel linking the outpost's various inflatable habitats and laboratories. Hadley dumped the contents of her backpack onto the bed and tried her room's screen system. She tapped a few keys, including a callout labeled GUEST: GIBSON VAN CLIVE, and began sorting the jumble on her bed. She turned at the sound of a rap on the open door.

"Hey," Gibson said, leaning against the low doorframe. "Not bad for an inflatable, huh?"

"It's heaven after that cruise. Where'd they put you?"

"Two doors down on the left. You called?"

He entered the room and sat on a chair next to the bed. The automatic door slid shut. Overhead lighting was tasteful rather than industrial, and there was even a subtle floral pattern on one wall.

Hadley sat on the bed and fixed her gaze squarely on Gibson, waiting.

"What's up? Out with it," he said.

"You know me well."

"Too well. You're the sister I was relieved to never have. So what's on your mind?"

Hadley steepled her fingers, squinted at the door as if to make sure no one was listening, then faced Gibson again. "Did you notice Joel? Since we arrived?"

"Reminds me of a caged cougar."

Hadley nodded. "He made the strangest comment."

"For Joel, that's saying something."

She slapped him affectionately on his bald spot. "Seriously. He pulled me aside as soon as we were off common radio and asked if we could meet. Tonight. At the drill rig entrance."

"Why there?" Concern braced his voice.

"Beats me, but he knows his way around this place. Maybe its private or something."

"He was here—what?—three years ago?"

"Something like that. Anyway, I guess I just wanted someone to know."

Gibson spread his hands. "You want me to come with?"

She shook her head. In Europa's 1/8g, she could feel her bronze hair swimming around her cheeks as if she was under water. "Joel's spooked enough. I don't want to scare him off of whatever he wants to tell me."

"What's it all about?"

"All he said was that it had something to do with our fellow passengers."

Gibson shrugged. "You're a big girl. But just be careful, okay? I don't want to have to tell your mother, who—I will remind you—charged me with protecting you at all costs upon penalty of death, that you screwed up and … and …"

"And what? What could possibly happen in a crowded science station like this?"

He held up both hands, like a bank teller in a holdup. "Good point. Never mind."

"I love it when you get that way. Mom would be proud. How's the diary going?"

"Slow. Interesting. More on that later. Let's go."

"We just got here."

"Time for Tefia's Taliesin tour, remember?"

Hadley frowned. "No. I completely forgot Tefia's Taliesin Tour. Has a nice ring to it, though. I really want to put my socks in that drawer over there. The little compartments and closets open with such a cool sound."

"Socks later. Tour now."

_*_

Ted let the door seal behind him, and then glanced around the small quarters. He didn't unpack. Not right away. He felt as though if he kept his carryall intact, he might be able to change his mind and go home again. But out this far, the trip was one-way. Once a person started toward the outer Solar System, there was no turning back until the end of the loop.

Despite his nervousness, the man of faith looked forward to seeing another aspect of God's creation, a new horizon upon which to explore and marvel. Wherever he went, he saw the Creator's brush strokes. And so he sat at his little monitor, returning to the place that always calmed him. He scanned down his growing list of quotations.

Who laid its cornerstone while the morning stars sang together? Can you bind the beautiful Pleiades? Can you loose the cords of Orion? Do you know the laws of the heavens?

Ted loved to look out the window at the sky, diamond dust scattered across a velvet field, and contemplate the words. And there were so many more.

He wondered what the others would think of such things. To him, nature was confirmation, encouragement, inspiration. But in the wake of The War, religious ideas were not popular dinnertime conversation. Then again, when had the carpenter from Nazareth been interested in popularity?

III

Gibson spotted Ted stepping out into the hallway as the rest of the fellow passengers paced, waiting. Tefia called from down the hall in that lovely alto voice of hers. "Okay everybody. Stop swarming like a bunch of bees. Time to get your bearings. Follow me."

Tefia walked them along the corridor. The large tunnel formed the spine of the science station complex, a long hallway with hatches on either side. The group passed two full galleys, a small coffee shop, a recreation room, storage facilities, and airlocks leading to various garages and exterior warehouses. The laboratories clustered on the west end of the assembly. To the east lay the living and recreational quarters. A central hub served as communications, a medical lab, meetings, and for administrative roles. An overhead skylight provided a nice view of Jupiter, ever-present in the Sun-drenched, ebony sky. Just now, Jupiter's violent moon Io was coasting in front of the giant crescent planet, dragging a trail of sodium in its wake. Gibson turned to Hadley. "Io's looking very leaky. I'll bet you'd love to see a couple of those gushers up close. Now *those* are what I call *volcanoes*."

"Even with lead underwear, you'd only get a few seconds before you fried." She peered at the pizza moon out there and muttered, "But what a few seconds! A volcanologist's dream. For a moment."

Tefia unsealed a large access port ahead. "And for your specific team, the new babies are in here. Just don't play with your toys until I'm not watching. All these people rummaging around all this delicate equipment—it gives me hives just thinking about it. And obviously, no food in here, please."

Tefia stepped through, shook her head, and grabbed a coffee mug from the shelf. "We make an exception for Dr. Cann. Jeff can't go anywhere without his coffee."

Gibson had to lean over to fit through the door. He felt sorry for Ted. The entrance would probably mean another bruise on the man's scalp. Beyond the hatch, the space opened up into a brightly lit cleanroom, with grated floor and vents along the walls. Air was flowing through the room at a gentle breeze. On the far wall, a long shelf held his attention. There, a parade of gadgets sat, waiting to be set free in the oceans of Europa.

"Ooh, ROVs," Dakota gushed. Aaron gifted her with a baffled expression.

"Remote operating vehicles," she explained. "Biobots. Little robots based on biological forms."

She was right: several looked like eels with whiskers extending out the front and rear. One stood atop spindly legs, each finned for swimming or walking. Another had a distinctly predatory appearance, with inverted dorsal fin and skids on the top for travel along the underside of the ice. Gibson bent close.

"I should have known you'd go for the one that looks like a shark," Dakota told him. "Such a boy." She did a double-take and leaned toward the biobot. "Ah, nice sampling devices. Filters for microbe study—"

"If such exist," Gibson put in.

"If such exist, yes," Amanda agreed, coming up to look. "And very nice little remote monitoring probes you can stick into nooks and crannies for any bigger beasties."

"Like whales?" Sterling asked, smirking.

"If only," Dakota said.

Amanda's eyes had the glow of a zealot. "These will do nicely."

"Sweet," Sterling Ewing-Rhys enthused, tapping the counter next to a streamlined, meter-long device. "Biomimetic robot with motorized fin."

Dakota studied the fish-like shape, the sleek instrument pod on the front, and the single fin running the length of the ROV's belly. "Why only one fin?"

"That's all it needs." Sterling pointed to the belly of the little craft. "It's weighted to stay upright. Designed to emulate the glass knifefish, if memory serves. These others are based on tuna, lamprey, salamander … nature's designs are often best." His voice trailed off. He turned toward Tefia. "Where are the comms?"

"We've got communications in a separate room for safety," Tefia said. "Headsets and cranial implants are next door. You can see that later. Jeff Cann will give you a briefing tomorrow. They're doing some kind of testing right now."

"I want one of these." Hadley had moved to the opposite end of the room. Behind a stack of crates, hanging from a low crane, stood a cherry-red craft. It looked like a torpedo with a glass nose. "Is this it?"

"Our little submersible," Ted Taaroserro said. "It seats two. You'll have to share."

"Happy to." She frowned at the compact passenger section. A doughnut-shaped outer ring protected the plexiglass cockpit windows from impact. "On second thought, I'm not climbing into this thing until it's time to find a volcano or two."

"You'll get your crack at the sea floor soon enough," Gibson said. "I've got a prelim schedule already. Just needs tweaking."

Hadley gazed at the machine. "It's so sleek. So skinny!"

"The passengers lie down in it," Ted said.

"Ah, I see now."

"They really missed a bet," Amanda said. "Shoulda been yellow. Then we would all have our own yellow submarine. Like the song?" She was met with blank stares. "Don't they teach the classics anymore?"

Tefia stepped around the submersible and opened another hatch in the wall. "And this, ladies and gentlemen, is the way to the pit. We can't get too close without suits. Safety protocol."

The group huddled at the window of the small antechamber. Outside, at the bottom of a gentle slope half a football-field away, a dome arced 10 meters into the sky. Engineers had linked the outpost to the distant dome by a sealed inflatable passageway. The dome's clear sides were fogged, misted over by the atmosphere inside. But within that dome, even at this distance, Hadley could just make out a fuzzy oval blotting out some of the ice floor, a dark maw, a window leading to Europa's abyss below. It was intriguing, impossibly dark. And it was chilling.

As Hadley looked toward the new dome, she noticed Ted and Dakota huddled against a side window, looking south. Remnants of another dome encircled a flat spot in the ice, a place once tunneled but now filled in with concrete-hard water. A spider web of cracks, ghosted lines frozen into the surface, fanned out from the smooth pond. Several warning cones stood guard at the foundation's borders. "That's where it must have happened." Ted sounded mesmerized.

Dakota shushed him. "Joel's right over there." The two walked on, but Hadley was drawn to the glass. Out there stood the image of doom, the visage of a disastrously failed mission, of the dark side of exploration. It was a place of death in the cold, one in a long line. Amundsen, Scott, Mallory. And, of course, Nobile and his red tent. She could hear the screams of the men, clinging to the envelope of the torn dirigible, carried away on a gale of arctic wind as the rest of their crew watched helplessly from the frigid landscape below. Stranded on the ice. Branded upon the historical record, not with heat but with a burning cold.

She felt a hand on her shoulder, and a gentle squeeze.

"Had," Gibson said softly, "Let it go. There's nothing out there that you need to focus on. This place is a second chance for what those people were here for, what they stood for." He nodded toward the ruins out on the ice. "Here, now, we have exciting things to do. Things to set right. Yes? Just walk away."

She nodded, wordlessly. She turned slowly from the window and rejoined the group.

Tefia's tour-guide banter split the cool air. "So for those of you who don't know, the pressure dome keeps the ocean water from boiling out, which it likes to do in a vacuum. The dome down at the Sidon Flexus site—800 klicks south of here—is similar in design. That's where some of you lucky ducks will get to go down. As for me, I'm happy to watch from the comfort of my warm monitor."

"The first humans in a Europan ocean," Ted said quietly. Everyone fell silent.

IV

Genevive's data suggests that we are within forty meters... excitem... I like her a lot. She reminds me of some French cooking show host. But I digre... adiation levels bad today, so we took most of the afternoon off. The submarine stands ready, dome fully pressurized at 0.35 bar, enough to hold down any water, we think. Fingers cr...

Still worried about Alison and deep space narcosis. She has the shakes and seems disoriented at times. No seizures or halluc... We can't connect very well with T's medlab at this distance. Too much interference. Of course, it will all be academic should... anth... meks.

It was one of the last entries, written sometime between March 12th and March 18th a decade ago. Gibson knew that the general events to follow were largely a mystery. The details might be important. That's what Hadley had said when she tasked him with this little assignment. It was all so clandestine. Hadley didn't want the powers that be to know she was spending any of her precious grant-funded energy on past history, and she didn't want anyone else to know, either. She had her reasons.

He thought back to the moment, in the darkened room, when she had broached the subject. Hadley shook her head, her coppery hair enhanced in the warm, dim light. "Some of these guys like DelGenio insist that it was simply too early to try a Europa submersible. They say it was just beyond the technology of the time."

"And you aren't buying it," he remembered saying.

"People can do amazing things with pretty rudimentary technology. The Polynesians sailed for thousands of miles on open seas in nothing but reed boats and dugouts. We sent people to the Moon decades before quantum computers. Usually, it's a failure of management and planning, not a failure of hardware."

"And what if the answers really are in there?" he had asked, tapping the battered electronic diary.

"Might be helpful to us. Any edge is a good thing in a remote place like Europa."

And so the path had been set. He needed to go back into the earlier files. Still, his eyes hurt and he was distracted. He dug the heels of his hands into his face, stretched his neck from side to side, and peered out the little window.

He always found it difficult, telling Hadley no. She had done so much for him, been there for him. He thought back to a party they had attended while he was a professor at the University of Arizona. The room was full of bankers and economists and political types and human rights advocates and a few artists. One of the guests, a man who was sober even this late in the evening, sauntered over to Gibson and Hadley with a small entourage of what looked like groupies. He looked Gibson up and down as if he was searching for lice. A smirk bubbled to the surface and he tipped his glass toward Gibson. *So,* he had said, *what do you do?* Gibson introduced himself and replied, *I'm a professor at the U of A.* The well-dressed man did not offer his name, but rather winked at the couple standing next to him as if to say, *Watch this.* His tone casual, he asked, *What do you make?* The question seemed disjointed from everything that had come before it, but Hadley recognized the non sequitur right away. She stood a little straighter, looked the gentleman in the eye, and said, *You know what he makes? He makes those people who will inherit our messes into the kind of people who will be able to clean up after us. He makes students believe they can do anything, and then he makes them go out into the world and do it. In short, he makes a difference. Now, why don't you share with everyone what you make? Or maybe next time you can make some silence before you open your mouth and say something embarrassing.* It had been a triumphal evening. Now, he knew, his adolescent friend had become his ally in adulthood.

_*_

Hadley opened her door and ushered Gibson in. He didn't sit on the chair she offered. He fretted. "I'm making headway decrypting that diary, even with all the corrupted data. All that's fine, but I really don't like this whole thing with you running off to powwow with Joel. Did you see the size of that dome? It's way out there on the end of a scrawny little tunnel. Completely removed from civilization."

"You call this civilization?"

He ignored the comment. "Why can't you guys just chat in the galley?"

"Ours is not to reason why. I think he's just nervous. What have you found so far?"

"A whole lot of bad data and a few interesting tidbits. Here, have a look." Gibson plugged a microdrive into the room's data port. The screen on Hadley's desk flashed to life, covered in text. Gibson scrolled. "There."

In a pinch, we can use Mars as an emergency relay, although one-way light time is about 33 minutes and getting wor ... ut what do I know? The jerk.

"That's not the only negative reference to a fellow crew member."

"Are you thinking the team just self-destructed? Seems a bit farfetched."

"I don't think their demise was that simple, but it might have been a factor. This next one is what concerns me." He tapped the keyboard.

... diation is wicked. It's wreaking havoc with all of our motherboards and some of our hardware, too. We really didn't expect this much damage. All the computers are getting confused, but at different times. It's been near impossible to predict which little brain is going to lose its mind when. The main rover is out, at least for n ... dary rover still moderately normal, plus our two short-range minirovers. But our little sub is another story. Mendelson needed to swap out an entire quantum memory plank and then gr ...

Hadley sat back. "Radiation damage. There it is again."

"Could be. Back then, who knows what kind of systems they had? Maybe everything just fell apart around their ankles."

Hadley shook her head. "I don't think so. They brought three or four of just about everything."

"Sure, but not full-on computer systems." He rubbed his eyes. "I don't know. I'll keep digging."

"Dinner?"

"Sure. But I'm not very hungry," he groused.

"Sweet. The big, protective brother I never had." Hadley patted his back. She didn't say it, but the thought of her upcoming meeting had dashed her appetite, too.

_*_

Gibson and Hadley entered the galley to the sound of raised voices. The expedition members stood in a semi-circle with Dakota in the middle. Dakota was leaning over a table toward a tech from Taliesin, her hackles up. The man—who was wielding a hypermop—wore a sheepish look, and his shoulders sagged. His eyes were sunken within pools of laugh lines, but there was no laughter within them, just an ancient sadness. He held his left arm close, as if it didn't work quite right. He looked vaguely familiar to Hadley. Perhaps he simply had one of those faces marked with the universality of human toil.

The lines in Dakota's neck stood out as she bellowed, "A waste of money?"

The man brushed frazzled strands of gray hair across his barren scalp. "It's just that people have been digging around the ice here for twenty, twenty-five years, and nobody's found any frozen beasties, not even little ones. How do you explain that if you have an ocean full of critters?"

Dakota pointed toward the window. "A lot of those brown blemishes out there mark places where salts have come up from below. And yes, that means there's an exchange between sea and the ice crust. But that doesn't mean you would bring up Humpback whales and walruses in the migrating ice. What it does mean is that big stuff gets filtered out and the fine stuff surfaces."

"My point exactly—" the man began, but Dakota held up her hand.

"But…what *that* means is that the organic material that *does* survive the trip up is delicate enough to fall apart in the surface radiation. Even if something as big as a microbe came up, it would rarely be preserved, as all the searches attest. We have to get inside, down into the ocean below, to answer the ultimate question of life on Europa."

The man shrugged and mumbled something about caves, trying to put some distance between himself and the table.

"The ice caves are our best bet for finding something, and my bet is that someone will one of these days. Just because they haven't doesn't mean there's nothing down there. We've found lots of complex organic chains, and they've got to be fairly fresh to have survived."

The man gripped his mop handle so hard that the white knuckles stood out against the crimson skin of his hand. "Maybe so. But I'll bet you're after something bigger, aren't you? The white whale. Yes, that's it. Europa's own Ahab in search of the beast. But you should be careful. We all know how Ahab's last scene played out."

"Buddy, I don't know what axe you've got to grind, but you're the one who should be careful."

"Gee, look at the time," he said, not even glancing at his wrist monitor. "Break time. Bye." He left at a remarkable pace for a man his age, waving behind him.

"That guy likes to speak out of complete ignorance," she sulked.

Hadley put her hand on Dakota's shoulder. "That's okay. It was a teachable moment. You gave him something to think about."

"He was casting aspersions at biologists in general."

"That's all right, dear," Amanda put in. "I'll bet you and I have higher salaries than he does."

"At least for this grant," Dakota said. Everyone laughed. She looked at Hadley self-consciously. "I actually never read it."

"Read what? The grant?"

"*Moby Dick*. What was Ahab's last scene?"

Hadley pressed the palms of her hands together. "Well, in the end, Ahab hunts down his great white whale for revenge, but the whale has the last word. Ahab gets tangled in the line of his own harpoon, and Moby Dick drags him down to his death."

"Do you think I'm a fanatic?" Dakota asked quietly.

Hadley smiled and patted Dakota's shoulder. "I think you're a good researcher, and I think we all need a few calories in our systems."

Dinner passed slowly, despite the celebratory mood of a team freed from an interplanetary cruiser. Joel was the only AWOL crewmember at dinner.

"He's mercurial," Sterling said.

Amanda pointed a fork at him. "Mercurial? How literary! I didn't think they let you robotics dweebs read anything but quantum circuit planks."

"Somebody forced me to read in college." His easy southern inflection pulled the words into a graceful train. "It wasn't pretty, but some of it stuck. Know what else I read about? Umberto Nobile. Hadley, there's a family tie-in, is there not?"

Gibson stole a nervous glance at Hadley. She tried not to roll her eyes in anguish. *Here we go again.* "Good ol' Umberto was the great uncle of my great grandfather, Giacomo."

"You don't look very Italian, dear," Amanda interjected, tapping her plate with the handle of her fork.

"Hey, Christopher Columbus had red hair."

Sterling looked skeptical. "So, you're following in your descendants' footsteps?"

"Ancestors. Yes, to a point. But I'm more interested in what's underground than what's above. That Nobile liked to dogsled and fly dirigibles. I prefer to look down the throat of a good volcano." She said the last with gusto.

"Here's a useless party fact," Sterling said. "Nobile's dirigible was almost exactly as long as our cruise ship, if you don't count the ion engine scaffolding in the back. I looked it up."

"Yeah, but it was mostly hot air," Amanda said, looking at Aaron, "like some people I know."

"Gas," Hadley said. "Hydrogen. Now if you had said methane…" She looked at Aaron with a smile. The group broke into laughter. Aaron remained dour.

Sterling placed his fork on his plate delicately. "I do admire those guys. The spirit of exploration and such."

"Man against the elements," Orri added. He sent a sideways glance at Hadley. "Or woman. Sorry chief."

"Golden Age of Exploration, they called it," Dirk offered. "That's partly how I got into piloting. Some people look at my job as interplanetary bus driving. Not me. I see all this, all that you do and the small contribution I make, all that as keeping it going, sort of a continuation of that exploring push."

"Gallant," Sterling said. "But then again, some of them just didn't know when to give up."

Dirk looked puzzled.

"Hadley's relative, for example," Sterling said, "had the first flight over the North Pole in his dirigible."

"The *Norge*," Hadley said, pushing the tepid food around on her plate.

Sterling nodded ascent. "The *Norge*, yes. But then, just a few years later, he set out again in another big balloon, the *Italia*, I think it was."

Gibson leaned back in his chair, wiping the corner of his mouth with a napkin. "Well, yeah, but he had a reason. Some people argued that he hadn't really made it over the North Pole and he wanted to do it again."

"Byrd even claimed to have done it before anybody else," Sterling said.

"Questionable claim," Dirk chewed. "Probably not."

"So what was the big deal with the *Italia*?" Amanda asked. Everyone looked at Hadley. She was silent. She didn't want to go through it all again. She remained in denial of the fact that she probably would have to field such discussions for the rest of her natural life.

Gibson filled the pause. "The flight was a disaster. The expedition definitely made the North Pole, but on their way back they crashed on the ice."

"Seems like you could walk away from a balloon crash," Amanda quipped.

Gibson shook his head. "Not this one. A gust drove the dirigible into the ground and tore off the gondola. Six men were stranded in the envelope as it blew away. They never found it. Nobile and his gang were stranded on the ice, but they had some supplies and a radio. Eventually most of them were rescued."

"But not before others were lost," Sterling said, punctuating each word with a slight pause. His speech quickened. "This is what I find so crazy about those days. Arctic expeditions are just full of irony, missed chances, weird coincidences. Scott died just kilometers from a food cache. Where Scott used ponies, Amundsen chose sledges like the Inuits use, and he made the right choice. He lived. Scott died."

"Or Shackleton," Dirk offered. "He and his crew kept missing rescuers, just barely, because of unlucky weather and so forth, but they made it home after all. Roll of the dice."

"And bravery and ingenuity," Sterling continued. "And on this one, while people were out searching for Nobile and his crew, the great explorer Amundsen disappeared while looking for them. After all he'd been through. Irony, friends, irony."

"Didn't Nobile abandon his crew or something?" Aaron asked.

Hadley bristled. "No. He *absolutely* did not. You see?" She was looking at Gibson. Gibs would come to her defense, wouldn't he?

Gibson lowered his tone into that of a lecturing professor gently correcting a wayward student. Turning to Aaron, he said, "He was accused of that, but that's not really what played out. This plane spotted the wreckage and the pilot was only able to take on one person. Nobile wanted him to take an injured mate, but the pilot insisted on taking Nobile himself, so that Nobile could coordinate the rescue effort. After they made it back, it took some time to get help to the others. But no, Nobile didn't abandon anyone."

"And he spent years trying to clear his name," Sterling added.

"Successfully, I might add," Hadley said, bitterness in her voice.

Aaron looked down at the table. After an awkward silence, with exchanged glances all around, Dakota cheerfully said, "Those were the days. Out here you get fried by radiation or frozen stiff or asphyxiated, but at least you don't starve to death. Not a good way to go." She shoveled a spoonful of linguini into her mouth with just the slightest smile.

Sterling leered toward Dakota's plate. "We can sure freeze here, but we've got tons of food. You gonna eat those Brussels sprouts, Dr. Barnes?"

She pursed her lips as if she'd just bit into a lemon. "Be my guest."

Hadley's thoughts played with the ghosts of those explorers who came before. She thought of Scott's heartbreaking last entries in his brave diary, found with his frozen corpse long after his expedition failed to turn up on time. She thought of Nobile's men, shivering within their red makeshift tent, waiting and waiting for help. And she remembered the modern researchers, the ones who had come just before The War, before them; of the dome ruins at Taliesin and those even farther south, at the site of the lost expedition; of the filled-in shaft frozen in the ice. How many bodies were still down there? And the diary that just might, with a little luck, cast some light on those last days at the southern station where they had sent the submarine crew to their doom.

They must be careful here. Very careful. Nothing must go wrong. She would not be an addition to a long line of lost explorers. And on her watch, no one else would, either.

She watched the time crawl by. Finally she excused herself. Gibson gave her a thumbs-up. Nobody seemed to notice.

V

The journey from the galley to the ice dome was labyrinthine and lonely. Hadley struggled to remember the afternoon's tour. She made her way past the darkened galleys, abandoned laboratories, and empty common rooms, and finally to the clean room. Most personal quarters had been closed up for the night.

At the entrance to the antechamber hung a row of yellow environment suits, looking like a squadron of deflated recruits. She dutifully donned one, putting on the helmet but leaving the visor up. She ducked through a hatch emblazoned with a sign that said,

Danger: Vacuum/Pressure Juncture Suits Required

She had noticed several of Taliesin's veterans coming and going through this hatch, many in coats rather than full suits. But the sign made her thankful she had taken the time for the suit. She opened the hatch, stepped through into the tunnel, and sealed it behind her. The tunnel was poorly lit, simply an inflated tube with some type of metal sleeve on the outside. It was cramped, claustrophobic, just as the gondola of the *Italia* must have been in past days.

Despite the layers of insulation in the tunnel's fabric, she could hear the crunch of ice beneath her feet. The walkway's sides undulated with her every step. Its thin walls sent goosebumps along her arms.

A small overhead light illuminated the beginning of a descent. Steps had been cut in the ice underneath the tunnel, making the climb downhill easier. After a dozen steps, Hadley found herself in another straight, level corridor. At the end of it, 10 meters away, was the entry to the dome.

As she approached the hatch on the far side, she could feel the continued drop in temperature. Her breath sent clouds of vapor drifting through the beams of her helmet lights. If she closed her visor, she would be warmer, but she needed to hear. She cranked the suit's heat and unsealed the door. Above the hatch was a darkened sign that read,

Drilling in Progress No Admittance

She would definitely say something about this cloak-and-dagger stuff that Joel insisted upon. It was wearing thin. Whatever the circumstances, their next meeting would have to be over a hot latte in the coffee shop. As she swung the door open, frigid air blasted across her face, into the neck of her suit, and down her back. She shivered violently, but didn't close the visor. Instead, she called out. The rotunda was well-lit by overhead lamps. Scaffolding hung over the dark hole like a menacing insect, cables snaking from it across the floor to some equipment on the far side.

"Hey Joel. You in here yet?"

Silence.

Great. I go to all this trouble to get stood up?

"Joel?"

The room wasn't big enough for someone to get lost. A mesh of thin cable netting surrounded the hole at the dome's center, enough to keep people out but not enough to hide the dark passageway to the underworld. The sides of the maw's vertical shaft were glassy smooth, polished by weeks—perhaps months—of drilling through Europa's crust. It lay open to the Stygian waters down in the darkness, an untouched, primordial ocean. What lurked beneath?

She circled it slowly, watching steam rise from the Hadean waters kilometers below. Surreal echoes from those waves swirled up through the shaft, sounding like distant, maniacal laughter. Thin ropes of ice trailed from the hole's edges, as if fresh water had been dribbled across the floor from inside the shaft. As she backed around the opening, she made her way past a pile of jumbled equipment. Her boot brushed against something. She knew immediately that it was something not mechanical, not artificial, but soft, flaccid. Her first thought was that something had climbed out of the shaft. She jumped, pivoted, and looked down on the body of what was once Joel Snelling. He was in a heavy coat, with no environment suit, dressed like a Taliesin pro. His face was bloated and his skin splotched and blistered.

The room leaned to the left, then lurched right. Hadley saw sparkling lights everywhere. Suddenly, she realized that she, too, was lying on the ice. Before her, less than a meter away, lay the pit. As her vision cleared, she pushed away from the terrifying shaft. She rolled over and came face to face with Joel. He didn't look any better than he had a moment ago. How long had she been out? She sat up and fought down the gorge in her throat. She keyed the microphone in the suit, hitting the emergency channel for the medlab. It was only a formality.

VI

"Freezer burns, essentially," Aaron declared. "Pretty simple. That's the cause of death. He should have worn a better coat."

Hadley shook her head. "I just can't believe that. Why would he come out here dressed like this?"

Tefia Santana had laid the body on a gurney. Her attempts at resuscitation, after those of Hadley's, had been to no avail. Now, she gazed at Aaron with thinly veiled skepticism.

"Hypothermia; interesting thought," her tenor guarded. "While it is a pressure-sensitive area, people do sometimes come in here without a bulky environment suit."

"Nope," Aaron said dismissively. "Seen it before. People don't prepare, and the cold brings them up short."

Hadley turned on Aaron. "Where the hell were you when I called a medical emergency? Your enthusiasm is less than impressive!"

He narrowed his eyes. Hadley was the head of this expedition, but his expression acknowledged no authority on her part. Throughout the preparations for the trip, and during the cruise out, his emotions had run the short gamut from flat to bored. Now, she saw something new: belligerence.

Shuffling footsteps announced the arrival of several more people. Gibson entered first, followed by Dakota and Sterling Ewing-Rhys. "We came as soon as we—" Gibson's voice trailed off. Dakota's eyes were red.

Aaron snorted and headed for the door. Amanda, who had been silent throughout the discussion, turned and left with him. Hadley scanned the faces of her colleagues. The expressions ranged from horror to sadness, all except for one. Sterling Ewing-Rhys' eyes were riveted to Joel's form. He had a look she couldn't quite place. It was almost a look of recognition, of something familiar seen out of context. He turned abruptly and left. Orri arrived, filling the doorway. Even before he saw the body, sadness etched his face.

"Joel had already done a tour here," Tefia said quietly. "He knew protocol. His dress doesn't necessarily mean anything."

Hadley peered at the corpse. "When I first saw him, he looked different. He had hives or blisters or something."

"I'll do up a blood panel, of course. It will take some time; the medical staff is stretched thin here. Sorry for your loss." Tefia pulled the white sheet over Joel's face. Hadley turned away. What had he wanted? Why all the secrecy? And more importantly, what would this do to the expedition? This was an inauspicious beginning.

4

Outside

I

"It was your fault. It *was* your fault." Tormenting voices echoed with lunatic laughter.

"Not on my watch!" came Hadley's reply. But her voice rang empty, too weak to be heard. "Not on my watch!" she yelled.

"But whose watch, then?"

"Take responsibility," said a third, scornfully. "Be accountable."

Umberto Nobile emerged from the darkness, a Rembrandt chiaroscuro. He covered his head with his oversized gloves. "Stop!" he cried out.

The voice awakened her. "Stop," Hadley said quietly, again, as she moved from the world of nightmares to the surreal world of a Europan outpost.

And of course it was not her fault. Joel had gotten into something, or he had been careless. Or crazy. There were stranger ways of committing suicide.

But he wanted to tell her something, didn't he? Something about the cruise out, or the fellow passengers. Something beyond the professional, beyond the norm of teammate conferences. Did someone say something to him in confidence, share a secret they shouldn't have? Did a fellow traveler tip their hand on some proprietary intimacy? For the observant, it wouldn't take much to notice: a too-long gaze, a slow blink, a twitch at the corner of the mouth could speak volumes. But had Joel been that observant? What had he known? *What had he found out?*

It was funny how paranoid one's thoughts could be at night, alone in the dark. When morning came, those fears and phobias dissipated like the fog on a sunny day. Outside her porthole, the Sun was finally rising. Illuminated

© Springer International Publishing Switzerland 2017
M. Carroll, *Europa's Lost Expedition*, Science and Fiction,
DOI 10.1007/978-3-319-43159-8_4

spots of boulder and hill formed a chain of pearls between Earth and sky. Along the brilliant horizon, one could see a mist-like glow. It began at the surface but brightened a little ways up, then faded abruptly to a rich purple before fading away into the sky. The rare vision betrayed the telltale sign of the interplay between Jupiter's sweeping radiation fields and the molecules in Europa's ice. Jupiter's magnetospheric barrage stripped away minute particles from the ice, sputtering material into space and eroding away the surface. The trailing hemisphere, west of Taliesin Outpost, got it the worst, and there was an odd band of irradiated territory along the equator even on this side, a few hundred kilometers to the north of them. Atomic hydrogen and oxygen and other things seeded the dark sky with their spectral colors, but only for a few minutes each dawn and dusk. Later in the day, the phenomenon faded to invisibility, a ghost of what was.

As the Sun rose ever higher, she could feel its warmth through the window. Its ascent into the heavens seemed sluggish. Europa's 85-hour day gave the unfolding morning a lackadaisical feel. She craned her neck, but couldn't quite see Jupiter from her window. It was too high, too far to the north. That was a shame. She enjoyed watching the aurorae play across the mighty planet's poles, as it did on the top and bottom of Io—an eerie incandescence at the poles. Sometimes you could see material being stripped away from Io's sodium cloud, making its way along that energetic highway, the flux tube, toward the planet itself.

Hadley had always found Io to be a spooky little world. The tiny moon generated ten times the power of the ship she came out on, more energy than a thousand nuclear power plants, and most of it crackled along Jupiter's magnetic field lines. Supercharged electrons and ions chained Jupiter's energy fields directly to several of her beloved Ionian volcanoes, channeling two trillion watts of pure, raw energy into Jupiter's polar regions, powering its spectacular aurorae. Like a cosmic puppeteer, Io tugged upon its flux tube to make the aurorae dance.

Despite it being out of view, Jupiter's tawny globe cast a warm pall into the gloomy hollows of Europa's morning. The shrinking shadows unveiled rusted blemishes in the ice, areas where organics had welled up from underneath. Europa's oceans were rich in alien, enigmatic chemistries, but no one had yet pinned down a source for all the reactions that were taking place in the abyss. Perhaps her expedition finally would. She was certain the submarine volcanoes contributed a critical part of the mix, but she was also sure Amanda and Dakota were longing for biological explanations.

She had let herself get lost in Europa's enchantments. Now, it was time for a shower, time to shake off the dirt and the night's doubts. Was this expedition cursed? Would she follow in the footsteps of Amundson and Scott? She felt the weight of responsibility, a heaviness where there was little gravity. Joel was gone, and she had a hole to fill in her crew manifest. Last-minute additions could be tricky, but Taliesin seemed to have more than its share of sharp experts. She had to find a submariner willing to make the long trek south. The obvious choice was Jeff Cann, but would he be available or even willing? More immediately, she had to oversee today's testing outside. This was going to be a long day.

_−*−_

Hadley headed toward the main control room. As she passed the lab, she heard voices, and she recognized Dakota's laughter. She peeked through the small window recessed in the door.

On the far side of the lab, Dakota sat with a joystick in her hand and a comms halo atop her rippling blonde hair. With each move of the control, one of the bots on the other side of the room made a slight noise—a servo working, a fin tipping, a probe retracting.

"That's it," said Jeff Cann, standing just behind her. He leaned over her shoulder. "But more like this."

Cann reached forward and gingerly guided her hand on the joystick. She let him.

Hadley felt a motherly smile cross her face. The kids were bonding. How sweet.

II

"Back home, the cold rages. It comes in gusts and blizzards. But here, it grabs hold of you like it's got talons. It sinks in, deep. It draws energy out of everything. Leaves you numb." Orri lifted a clenched hand like a claw, underlining his own thoughts. Then he crossed his eyes.

"You're sounding particularly morose today," Hadley scolded. "Maybe you should stay inside where it's cozy. Then again, those big ice floes you love so much are on the other side of that airlock, so it's going to be hard to study things. At least you'll be warm."

He smiled for the first time since Joel's "accident." "I suppose you're right," he said. His face fell. "Of course, it's not the ice that's giving me the chills." He glanced in the direction of the medlab, where Joel's body lay under a faceless sheet, in a freezer locker much warmer than it was outside. "Hadley, I—"

"Ready?" came a voice from the hall. Gibson leaned through the doorway into the room. Like Orri, he was decked out in an environment suit. He was wearing his helmet. "Hey, put your hat on. Or were you going to hold your breath?"

"That would defeat our purposes." The big man squared his shoulders and glared in the direction of the airlock. "I really hate EVA."

"Hey, it's sunny outside—always is here—and suit checks are important, boys," Hadley said. "Go get 'em."

–*–

Hadley sat in the control room with Tefia, monitoring the equipment checks, wondering what Orri had been about to tell her. Gibson was at the long-range gliders, but—like a mother hen—always in visual and radio contact with his EVA partner. Orri, still just outside the south airlock, was putting his suit through a battery of tests, along with several small all terrain vehicles. The south airlock seemed a prudent place to check out the expedition equipment, as it got far less ingress and egress traffic than the central airlock.

"Rad levels look good, guys," Hadley radioed. "Ted says we're good to go for a while, at least."

"Roger that," Gibson said, heading toward a glider. Already, the flatbed had been loaded with equipment and supplies for their journey south. Hadley watched his camera feed as he stepped up into the airlock at the front passenger pod and cycled it shut. "Entering our Glider One now," his voice crackled.

"Soon as you can, switch to the internal comms, Gibs. Your connection is lousy from inside there."

"It's the shielding on the cab," Tefia interjected.

"No prob. I'm unsealing my helmet and plugging in now."

Gibson's feed went blank. His face appeared from another angle as the glider's system kicked in. "Power levels look perfect. Got a hot little nuke back there just waiting to take us to our geyser wonderland."

"From your mouth to God's ears," Hadley said. "Orri, how's tricks?"

"Just checked out the second ATV. On to the third. They're really fun in this low gravity. BMX would be a blast on this stuff."

"Yep," Gibson said, sounding like a kid. "Off-roading is all we'll be doing out here. I can hardly wait."

Tefia kept her eye on the two medical readouts. Hadley watched over the fluctuating radiation levels, still within manageable limits for their advanced shielding. She thought about where those ATVs and the behemoth gliders were going to take them. She wondered what was waiting for them. What would an active Europan geyser look like up close? Would they feel it thunder beneath their feet? Would it cast shimmering spray across the landscape and their suits? And would they, after all the fireworks, be able to actually venture into the pristine ocean of Jupiter's smallest Galilean satellite? She wondered how far the others had gotten. She shivered at the realization that some of them still might be under the ice where she would be walking.

As in any test session, frenetic activity periodically gave way to interludes of abject boredom. Hadley looked over at Tefia. "So these gliders, I'm not really checked out on one yet. What's the deal with them?"

Tefia sat straighter in her chair. "Carry a couple tons of stuff. Pressurized cabin at the front. Obviously there's nothing for them to actually 'glide' on in a vacuum. They're just like a Moon bus, a sort of hydrofoil for the airless environment. Thrusters keep them up and they skim along the ground. Pretty sweet. The ones you'll be using can cruise at 95 km/h."

"Not bad, not bad."

"Still a long drive, I imagine. What's your route?"

Hadley squinted, remembering the details. "Current plan is to head south-west past Sarpedon Linea. We'll skirt Balor Crater around the south rim, and then cross Adonis Linea. As soon as we're far enough south, we'll cut straight across, into the borderlands of Thynea linea and arrive at the Sidon Flexus dome outpost from the east."

"The engineers have set things up nicely for you all, mostly telerobotically from orbit. I think our personnel only had to make one trip down in person, to make sure the dome was secured to the ice and the bore was drilled to depth. They didn't stay long. Not the explorer types. Everything is in good shape and waiting for you down there. But the dome outpost is still a couple hundred klicks from the geyser activity, isn't it?"

Hadley swallowed a sip of coffee. "'Bout one-fifty. We stay at the dome until the last minute, and then we'll make a mad dash for the geyser site at one specific arc on Sidon Flexus."

Tefia jabbed a pen in the air at Hadley. "Smart, keeping out of the high radiation until the last."

Hadley nodded vigorously. "That's the idea. Ted's idea, actually. He'll be monitoring the radiation levels, auroral activity on Jupiter, and the like. We'll leave telerobotics on site at the geysers so we can continue to monitor from the safety of the dome. We'll only be in the hot zone for a couple days, max."

"Makes sense. And then you spend the rest of your time bobbing for Europan apples?"

Hadley laughed. "My biologists would love to find apples, I'm sure. But I'm holding out for a Europan Vesuvius, down on the ocean floor."

"Hot times, girl. Just remember: I don't want you guys out longer than 500 hours in total. It's too far, too dangerous. You'll be adding rads constantly, and there are health guidelines on such things. Three weeks. No more."

"No more," Hadley agreed.

Ted checked in each half hour, giving updates on Jovian aurorae and what they might mean for upcoming radiation levels. His estimates matched those of Tefia. The energetic environment remained steady.

At lunchtime, Orri's voice crackled over comms. "Hey Hadley, can you switch to private channel for a second? I want to check that frequency on this suit."

"Moving now," she said, looking at Tefia doubtfully.

Tefia glanced from side to side and said, "I think he wants something. Time for the little girls' room. Be back soon."

"Thanks," Hadley said. As soon as Tefia left, she keyed the microphone. "We have privacy. Do you read? I thought we already checked this."

"Actually, we did," he said awkwardly.

"What's up?"

"Do we have, er, relative seclusion?" He sounded nervous.

"Just you and me, babe. Are you okay?"

"Fine. But I think I figured something out. At breakfast a couple days before we got off the transport. Something someone said. I think it's important."

"Give it up, Orri. The suspense is killing me."

"Not on a potentially open channel. I'll come by your quarters tonight after dinner. Good?"

"I'll count the minutes, Mr. Melodrama."

"Switching back to common feed." Orri said it in a businesslike voice.

Gibson and Orri spent a total of nearly ten hours in the Europan sunlight. Jupiter floated above, a giant scythe in the sky. The two barely had time to check the suits of all the expedition members, cycling each through a battery of tests in the hostile Jovian environment. Finally, they climbed back through Taliesin's south airlock.

"You guys look like wilted cabbage," Hadley said in greeting.

"And you smell like cioppino," Tefia added, pinching her nose dramatically.

Gibson handed Hadley his helmet. His sparse hair looked painted on to his forehead. His face glistened in a sheen of moisture. "Well, thanks. While you guys were sipping your coffee in your comfy chairs—"

Hadley held up a hand. "Save it, princess."

"You both feeling all right?" Tefia asked.

"Tired, but good," Gibson said.

"Tired is right." Orri shook his head, a bead of sweat rocking back and forth on the tip of his nose. He stood in stocking feet and thermal undergarments. His suit lay like a deflated mime in the corner of the suit-up corridor. "But it's going to be a beautiful trip. Being out in it for a while gives you an entirely different sense than you get staring through a window. Here, Gibs,

I'll take that." He reached over and took Gibson's suit with one muscular arm. Orri turned and started back down the corridor toward the suit-up area.

"Thanks, man," Gibson said. He stretched, sniffed, and said, "I'm hitting the shower." He raised his voice in his partner's direction. "I'm fine, but Orri needs one." Orri grumbled and continued on. Gibson grinned mischievously. "See you all in a bit."

Orri had disappeared down the corridor. Tefia looked at Hadley. "All in all, I'd say we've had quite a successful day."

"Productive," Hadley added. In many ways, she thought. She headed for her room, feeling on top of the worlds.

III

Hadley sat beside Gibson at her desk, watching the monitor and keeping an ear out for Orri. Text waterfalled down the screen. In the darkened room, the monitor's light cast an eerie blue glow on their faces. "Were they at the Adonis Linea site at this point in the text?" She stabbed a finger at the monitor.

"Yeah, at their submarine entry point, but I need to scroll back a bit. Have a look at what Culpepper says here."

My radiation readings have been pretty scary since we left; certainly higher than I anticip... heard a couple ghouls in the Taliesin galley discussing how many of us they thought would make it back. I didn't have the heart to tell the others, but it's plaguing me. I especially didn't want Genevieve to hear. She seems so delicate, even though she's a big woman and obviously self-assured, at least when it... aybe I'm just smitten with that accent. As for our other female team member, the good Dr. Sylven is the original ice queen. Granted, Alison is not here to be social, but give me a br... w what her problem is. Going this deep into sp... an be tricky psychol... chiatrist said we should watch for. So I'm watching. I'm watching everyone. And I asked G... atch me.

"Now, this stuff," he said, "comes from the next day as far as I can tell."

All the radio interference has me a bit worried. I expected it, of course. But we'll be a long way from help, and I'm counting on the relay stations we'll be setting up to keep in touch with Taliesin. The radio-environment here is quite complex, what with Jupiter's magnetosphere, Europa's induced magnetic fields from the oceans, our own... high frequen... In a pinch, we can use Mars as an emergency relay, although one-way light time is about 33 minutes and getting wor... ut what do I know? The jerk.

Gibson tapped on the screen. "You already saw this bit here." The text began to scroll again, reversing back in time. Much of it was unreadable. "I haven't gotten much out of this section."

"Were they successful at setting up all the relays?" Hadley asked.

"Yes, and they actually worked, at least for a while. We know Taliesin carried out several comms sessions on their trip down using the relays, and they had a few successful ones once they set up the south base and drilling started. Apparently, there were several more routed through Mars as the relays began to fail."

"Three, that's right," Hadley said. A look of panic shadowed her face for just a moment.

"How did you ever find that out?"

She shrugged. "You're not the only one doing research."

Gibson let it go. "What I'm wondering is what kind of physical and psychological workup they had on Alison Sylven before the mission. She may have been a loose cannon."

"No good," Hadley said. "All those records were lost in The War. The memory worm that wiped out all the government cyberstructures also took out most of the science stations, even the remote ones."

"Collateral damage," Gibson mused.

"Not so innocent, I would guess. The Eastern forces really didn't like what science was doing. They were phobic."

"Can't blame them completely. Science in those days wasn't all so good. Look at the eco-disasters, the genetic experiments, the trigallium fallout."

"True enough," Hadley said. "I always found it ironic that the Eastern alliance counted the Seven Sisters among their ranks. And for those girls, science was their weapon."

"Medical science, if you can call what they did *science*. I'd say science was more like their paintbrush, with a splash of torture on the side, and they were very dark artists. Ah, here it is."

"Wow, that's a big chunk of type," Hadley said, peeping over the rim of her coffee cup. "You've been a busy little bee."

… ys he's all impressed by the care they are taking with the technology, the batt … s all well and good that he's so happy, but I'm not sure just how impressed I am. Just yesterday I watched Donnie do a really half-a … Those things are life-and-death, if you ask me. And with only a couple days before we head south, I'm not real thrilled.

Mendelson is definitely a feather in the cap of our little troupe. Talk about world class! And a sweet guy, too. We talked for an hour about the latest World Cup and the possible parallel league on Mars. Very controv … lson told me about living half a thousand feet below the Antarctic ice for a couple months. Guess a little trip to Europa won't bother the guy. Clearly, after hav… eard the stories, the guy can drive, even if it's under water.

Gibson leaned away from the screen. "Dave Culpepper didn't seem to think much of Donnie Ramirez."

"At least we know he liked Mendelson," Hadley observed.

"You've got to admit, the guy was a pretty big deal."

"Yeah, I always wondered what he was really like, the way you wonder about Einstein or Hawking or DeGenerick. Now…"

They both leaned forward. Another paragraph blanketed the screen.

Tractor day. The two big rovers are ready, so they… treads instead o… nd frankly, I'm pretty much done being cooped up here. I say, let's hit the road, or the ice, or whate… secondary rover, but we'll have to make do. We don't need the third, I guess.

"So where does this leave us?" Gibson asked.

Hadley gazed out the window, but her focus was inside, within the words burning the screen. She pivoted to look at Gibson, splaying her hand and touching her index finger. "First, we know they were worried about DSN, as they should have been. Back in the day, with those long travel times, it wasn't as rare as it is now. Second, we know they went out with two rovers instead of the planned three, and they had two minirovers for local excursions."

"They weren't good for much else," Gibson interjected. "More like little three-wheeled ATVs."

"And we know they made it to the dome at Adonis and set up operations."

"And made it through the ice, presumably," Gibson said.

"More than presumably. They eventually deployed the sub." She was on her fourth finger. "And we assume one of the small rovers had some catastrophic failure as we've seen it from orbit in completely the wrong place. And we know communications degraded fairly rapidly toward the end." With this, she bent her thumb down.

He frowned. "But at that point, everyone seemed to be okay, at least as far as Taliesin knew, so they didn't get all hot about sending a search party right away."

"Right, tragically. And we know that the radiation levels surprised them, and their equipment kept taking hits."

"Especially their computer hardware," Gibson pointed out.

"Yes. Which could have contributed to an expedition-wide failure, or failures, of some kind."

"And don't forget the deep space narcosis."

"Sure, but was that real, or did Dave Culpepper have a good old-fashioned grudge against Sylven?" Hadley asked.

"Or a bit of both?" Gibson ran his fingers through his hair, still damp from his shower.

Hadley grimaced. "DSN can be a nasty thing. One bad apple can seriously ruin everybody's day."

"Or expedition."

Hadley's eyes glittered. "Exactly what I was thinking."

"Bottom line, we need more data," Gibson ventured. "I just hope I can squeeze it out of that poor little diary."

Hadley put her hand on his arm companionably. "Great job so far."

An alarm sounded. A recorded voice echoed down the halls.

MEDICAL EMERGENCY, SOUTH AIRLOCK

MEDICAL EMERGENCY, SOUTH AIRLOCK

Hadley's door chimed. She hit the open command on her screen. Tefia peeked in. "There's been another fatality."

5

Europa Asylum

I

"The lock was fully pressurized, but it's been cycled. Something must have gone wrong when he was outside." The tech sat on the floor, her face buried in her hands. "I never saw anything like this, ya know?"

Tefia Santana kept her hand on the woman's shoulder. "It's okay. I've seen a few dead people, and you never get used to it."

The tech looked up, her eyes rheumy. "Really? Even in your business?"

"If I ever got used to it, I think it would be time for me to hang up my med scanners."

Gibson and Hadley stood at the entry hatch of the airlock. Hadley crossed her arms over her chest and frowned. "This is bad. Poor Orri." She shook her head and looked up at the ceiling, fighting the tears at the corners of her eyes. "So they found him inside the airlock?"

The tech nodded. "Yes. It was sealed, but fully pressurized."

Hadley swiveled to Gibson. "Did you call the others?"

"Yep. It's like a rerun of Joel."

Hadley shivered.

"But this just doesn't make sense," Gibson scowled. "This is insane. He had no reason to go back outside. Both of us could hardly wait to get back in. We'd checked everything off the honey-do list. Why would he?"

"He might not have actually gone outside," Tefia said. "We'll check the feed, but his helmet wasn't sealed, just on loosely."

© Springer International Publishing Switzerland 2017
M. Carroll, *Europa's Lost Expedition*, Science and Fiction,
DOI 10.1007/978-3-319-43159-8_5

Aaron entered the cramped corridor and muscled his way to Orri's side. He examined the helmet on the floor, pulled the man's eyes open and flashed light into them. He felt for pulse. "Did you scan him?" he asked Tefia.

She nodded. "He's gone. I don't think it was asphyxia, but the suit data feed was turned off."

"That's odd," Gibson said.

"Maybe he wasn't planning on a full EVA. Maybe something distracted him." Hadley was grasping at straws.

Ted and Dakota arrived. There was no room for anyone else in the cramped lock, so they peered in, shoulders against the outside door frame. "This is just too sad," Dakota said. "They said this would be a dangerous trip. I guess they were right."

Amanda appeared.

Aaron stood up and brushed his knees off.

"It's clear that his suit failed—that's why we do these tests, after all—it decompressed partially, and the cold killed him."

Dakota began shaking her head, but the shaking made its way down her entire body. "Dear God, what's going on? First Joel, and now Orri. Two members out of a nine member expedition?" She looked at Hadley, then at Tefia, and then down to Aaron, who was watching her with concern. Her breathing sped up.

"It's like some curse-of-the-mummy thing. But who is the mummy?" She choked on the last word. Abruptly, she sat down on the floor. "I … I can't breathe."

Aaron stood and stepped over to her. "You're hyperventilating. Classic symptom of a panic reaction. Relax."

He looked down at her as if he pitied her, but he didn't offer any help. Tefia did.

"Nothing to be worried about, Dakota. Look at me. Look at my eyes. Now, breathe into this." She handed her a small plastic sack, a sample bag from one of the hanging suits. Dakota did as Tefia urged.

"That's it," Tefia encouraged. "Slow and steady. Better?"

"Better," Dakota said into the bag. She stood unsteadily. "Thanks."

"Why don't you all go someplace with a little more room and relax for a while."

"Splendid idea," Ted said, guiding Dakota by the waist. "I'll buy you an espresso."

"I was thinking other causes are possible as well," Tefia said quietly, aside to Hadley. No one else heard. Hadley grabbed her elbow and guided her down the hall, away from the Grants and Gibson.

"What other causes?" Hadley whispered.

"Dr. Grant may well be right. Death by decompression under the extreme conditions of this environment can present in a variety of forms. But this looks more like some kind of toxin to me."

Aaron came up from behind them, obviously eavesdropping. "Simple case of hypothermic dermal burns if I ever saw one, which means this outpost needs to reexamine its safety protocols. But you are the base physician, Doctor Santana." There was acid in his voice. "It's *your* report. Have it your way."

Aaron left histrionically. Amanda came up and put a bony hand on Hadley's shoulder. "He gets this way. Territorial or something. Especially when it comes to diagnoses. He'll come around soon. Don't mind his mood." She followed after her husband.

Tefia leaned toward Hadley and lowered her voice. "Look, I've seen hypothermia. Just look where I work. And I've seen symptoms similar to this. Not firsthand, but in the records. I really don't think your friend here died of natural causes."

"When you say toxins..." Hadley prodded.

"I mean poison. Accidental or administered, I simply cannot tell here. Something in the family of one of the c-DNA agents."

"That stuff they used in the labs, back before The War? Something outlawed on three planets?" Hadley asked.

Tefia quieted for a moment. "But the problem with that stuff is that it's really tough to recognize as an outside agent. Hard to detect chemically, so often the only recourse is to go by physical symptoms. And there's another problem: as Aaron is the second upper echelon doctor on site, and he has a different opinion, we can't do an autopsy. Against regulations. In a limited facility like this, it's considered an emergency procedure. For that, you need two signatures."

Hadley wondered if Aaron knew this. Surely he wouldn't stand in the way of further study, just to make sure Orri hadn't died of something that could be dangerous to the rest of the outpost. But the guy was twitchy, temperamental. Sometimes he seemed to stop cooperating with the team simply because he wasn't in the mood. "Isn't there another physician stationed at Taliesin?"

"We have a medic, but not of high enough rank. The rules are pretty clear. Unless there is direct cause for suspicion, with two primary physicians signing off on it, no autopsy until the body is back to a major facility."

"Does Ganymede count?"

"Yes, Port Dardanus. Mars is next closest."

"Cause for suspicion," Hadley muttered. "Orri goes back to the airlock, apparently pumps it down all by himself, dies in vacuum, and then pumps it up again with some kind of toxin in him. That's not suspicious?"

Tefia glanced up at the ceiling, and then looked down at Hadley, eyes intensely focused. "Had Dr. Sigurðsson been threatened recently? Did someone hold some kind of grudge?"

"No, not with Orri. Everybody loved the guy. He did say he had an issue to talk to me about tonight."

Tefia shook her head and looked back down at the late Orri Sigurðsson. "Not good enough. Could have been anything, from his paycheck to a personal issue with one of the other crew. They'd string me up if that's all we had. Sorry." Her eyes wouldn't leave the floor. "Still."

Hadley thought about her own words. Clearly, someone hadn't loved Orri. As Dakota had so ably pointed out, this was the second suspicious death within her small corps of discovery. And from a strictly practical standpoint, now she was missing a submarine expert and a glaciologist. That was not the best way to begin a scientific expedition into the treacherous wilds of an ice moon.

II

The rational part of Hadley told her that people die at Europa all the time, especially those untrained in the art of outside activity. What was it Sterling had said at dinner? *Arctic expeditions are just full of irony, missed chances, weird coincidences.* Perhaps. But the irrational part of her, the one that was sometimes right, warned her that two crew deaths were beyond coincidence.

She leaned over the table and lowered her voice, despite the fact that she and Gibson were alone in the galley. The smell of disinfectant and the acrid bite of burned food seasoned the air. Dishes clattered in the kitchen, serving as the only mood music.

"Gibs, you saw him. Did he look strange to you?"

"Dead people usually do."

"That's not what I mean. His skin. Did you see the blisters, the lesions? He had spots. They were faded by the time you got there, but he did. And you know what? So did Joel. But I was the only one to see those. And Tefia Santana said she was reminded of some toxic something-or-other. Something from the war times."

"You're suggesting Orri was poisoned?"

As soon as Gibson said it out loud, she flushed.

He laughed. "Do you know how crazy you sound? Not you, that is, but what you're saying. Why would anyone want to purposely harm Orri?"

The heat spread up her neck and into her ears. She began to sweat. "Lots of reasons, I suppose. Greed. Blackmail. A lot of people have baggage and dark histories from the war years. Or maybe it has something to do with someone's personal past."

Gibson straightened. "Look, I want to be supportive and all, but let's not get carried away. We are way out among 'em here, in an alien environment that will be trying its best to kill everybody at a moment's notice. We don't need to start seeing conspiracies." He looked at her and his bluster evaporated in an instant. Maybe he could see how embarrassed she was. Maybe he wasn't as sure as he wanted to sound. His tone softened. "Right, Had? I mean, maybe there was something, some contaminant, involved, and we should be careful. Watchful. I agree. Can we leave it at that level and not call a five-alarm crisis about some alien plague yet?"

"I wasn't planning on panicking." She stirred her tepid coffee too briskly.

He didn't look convinced. "Hadley, we've been through a bunch of things, a lot of history, and I trust you with my life and even my baseball card collection." She frowned. "Right? I'm on your side. I'm sorry if I seemed dismissive."

"No, no, you did *not*. You sounded apathetic. That's worse. Or like you thought I should be locked up."

Someone's laughter pealed down the corridor from the rec room.

Gibson grinned. "Maybe Taliesin is just the insane asylum you need." She didn't smile. He stood. "At any rate, get some sleep. G'night."

"Good night, Gibs."

She watched her old friend leave, wondering when he'd lost the starch in his spine. She looked toward the window. Sunset had draped its dark veil across Taliesin, the brilliant wilderness outside giving way to the velvet black one could only see in a true vacuum. Exhaustion took hold, and for the first night in ages, she was sure she would get a good night's sleep.

III

"Felicia Tanaka," Tefia said definitively, putting her coffee cup back on the tabletop. "She's your best bet, obviously."

Hadley relaxed her shoulders. With the loss of Orri Sigurðsson, perhaps that could work. She keyed in her latte order and took a seat next to Tefia. The coffee bar lacked the cafeteria ambience of the main galley, and in her opinion that was a good thing. "Could Taliesin spare her for the duration?"

"It's pretty damned important, wouldn't you say? Nobody's going to be mounting a backup expedition to see Europa's elusive geysers in the near future. You need someone with that kind of background, unless Gibson can fill in."

"He's got the knowledge. In fact, I'd say he was one of the leading experts on ice moon geology, but I just can't spare him in terms of the time. He's going to be busy enough as it is. My team selection was bare-bones, streamlined."

"Felicia's a good bet," Tefia repeated. "She's mapped several of the more obscure chaos regions, not to mention Thera and Thrace Macula. She was the first to show the complex plumbing through the ice between the two."

"That could be important, if we see the same kind of subsurface plumbing at our site. Do you think she would do it?"

"I'm all for the direct approach. Let's ask her." Tefia stepped over to the bar and punched a few buttons on a pad. "Dr. Tanaka, are you up for a drink? We need to talk to you."

Tanaka appeared remarkably quickly. She was thin and very short, below 5 feet tall, Hadley estimated. She carried herself with the grace of a gazelle. Like so many people working in low gravity, she kept her jet-black hair short, in a pageboy style. "Hello, ladies. What's the urgent issue?"

"Did I sound urgent?" Tefia asked.

"I could tell something was up."

Tanaka looked at Tefia with deep golden eyes, then at Hadley. Tefia glanced at Hadley expectantly.

"Can I get you a coffee?" Hadley offered. Tanaka shook her head.

Hadley folded her hands and grimaced. "Well, it's like this. Dr. Tanaka, how would you like to go on an exciting sightseeing trip? Be among the first to actually witness a Europan geyser firsthand? See some ice floes that only seven other people ever have?"

"Do all those people happen to be dead?"

Hadley took a quick slurp of coffee. "I was going to leave that part out." She swallowed hard.

"So you want me to replace your geologist?"

Tefia raised an eyebrow. "Word travels fast."

"What happened to him, anyway?" Tanaka's ebony pageboy swirled around her long neck like prairie grass around a willow.

Hadley cleared her throat. "An unfortunate accident."

"It's a long drive down there." Skepticism seasoned Tanaka's voice.

"Take us eight or ten hours on those hot-shot gliders."

"So just one day down?"

"To the submersible dome entry site. We set up camp there and then wait for the geysers. Once they become active and breach the surface, we shuttle over into the hot zone to observe up close."

Tanaka glowered. "You're talking a whole lot of radiation once you cross over into the leading hemisphere. I remember being in a cyclone in Thailand once. We got stuck outside, and if you stood against one side of this warehouse, you could see the rain machine-gunning everything around you, but you couldn't feel it. But if you stepped around to the windward side, you could drown just standing there. And that's what it's like here. Taliesin is nice and cozy. It would take a lot for me to step into the rain."

"Ah, but we have the best in umbrellas," Hadley grinned. "All the equipment is radiation hardened, with redundant systems carried in vaults for backup."

Tanaka tapped her chest. "Yeah, well, I'm not radiation hardened."

"Our suits are state-of-the-art, as are our countermeasure meds. And we won't actually be on site for more than a couple days. Then we dash back to the protection of our little dome outpost and venture through the ice to the ocean below. Wouldn't you love to see all those ice layers beneath Thynia Linea? All that structure in those mini-chaos ponds we'll be exploring? Chance of a lifetime."

Tanaka frowned, rubbing her chin between her thumb and finger. "That it would be." Was she weakening?

Tefia added, "I'm limiting them to 500 hours out. Radiation levels dictate about that."

Hadley tried to seal the deal. "The best part is that we're carrying a high yield reactor to generate energy fields around the outpost. Our dome will be shielded from most of the ambient radiation by its own little magnetosphere. It's a miniature version of what the cruise ships use on deep space trips."

"Fighting fire with fire," Tefia put in.

"Exactly," Hadley said. "Besides, the view from down south is spectacular. You'd be studying chaos and linea regions closer to the pole than anybody before you. And from there, we may be able to see Europa's faint aurorae. That would be amazing."

"Amazing." Tanaka's resolve seemed to be flagging.

"More than amazing," Hadley pushed. "Spectacular. Magical! And you'd be traveling with some of the greatest experts in all the worlds, some of the finest in history."

"There are some things that have been bothering me," Tanaka said. "The major chaos regions tend to congregate within about 40 degrees of the equator. There are exceptions, of course."

"So chaos regions congregate like geese flock and rhinos herd?" Hadley said, deadpan.

"English was never my strength. Cut me a little slack; it's my second language."

"So what's the bother?"

"We've found lots of salts and organics in those regions, which implies that they are getting at least some upwelling of ocean water from below."

"Makes sense."

"Especially if Europa has a thin crust of ice."

"Oh, I see. But you hold to a thick ice sheet?"

"My models would seem to indicate that the ice crust is not kilometers but tens of kilometers deep. Besides, the two biggest haloes on Europa—big, multi-ringed impact features—seem to have broken through at about 20 kilometers down."

"The big one's Pwyll, right?"

"Very good try. But Pwyll's a crater; didn't quite make it through to the ocean. I'm thinking of those large basins Callanish and Tyre. Great big bulls-eyes."

"So how do you get all that deep water to display on the surface?"

"Exactly my point," Tanaka said. "But there are ways. If you look at Europa's ocean as having convection cells moving warmer water from equator to pole, just like the rising water in a boiling pot, most chaos regions would probably end up where we see them. But if the ice is thick, it will also exhibit convection cells."

Hadley frowned.

Tanaka leaned forward slightly. "You see, you can have currents even in a solid medium like ice. Masses of warm ice will migrate through from bottom to top. They're called diapirs."

"And you'd like to find out which is the case."

"Lots of people would." Tanaka paused, then nodded, as if having come to a decision. Hadley held her breath.

"When do you need to know?"

Tefia said, "They leave in a few days, timed to the projected geyser activity. Soon would be good for Hadley, and for me as head of base operations."

"I'll let you know tomorrow." She said it with finality. She stood and turned in fluid motion, a bird floating on the air. She glanced back over her shoulder. "By the way, Jeff Cann told me he's going with you. That's a good thing: he was on the set-up team for the dome outpost, so he knows his way around down there. Besides, with Jeff in the mix, it will definitely be an entertaining outing." Silently, she vanished down the corridor.

Hadley grinned at Tefia. "So Jeff's on board. Nice of him to tell us."

"He gets around to things eventually. What happens if Tanaka says no?"

"We go anyway, and return with an overworked Gibson and a whole lot less data."

"Fingers crossed, then."

IV

Gibson studied the subsurface radar in Orri's files. He supposed it was his way of dealing with his friend's death, keeping it at arm's length through finishing the man's work. Without the big Icelander on the team, Gibson would need to pick up the slack, even if they did get Tanaka as a last-minute fill in.

Orri's detail map centered on their dive site at the edge of Thynia Linea, along the great Sidon Flexus fissure. Thynia sliced a 25-kilometer swath across the southern wilderness, its dark gray floor torn asunder along a prominent central ridge. Rows of pits on either side echoed each other, demonstrating that material was moving away from the central ridge in mirror image, much like the Atlantic seafloor spreading on Earth. Rugged parallel grooves and hummocks wandered along the length of the valley, offering a challenging route to any driver that might try to cross it. Gibson and his colleagues might be some of those drivers in a matter of days.

Gibson marveled at the detail of Orri's map. Nobody did it like Orri. The man loved detail, and his work always showed it. There was a sense of craftsmanship, of loving care, to his charts and diagrams. Could Tanaka do anything close to this? Could anyone? He wished he had known the big Icelander better.

Turning back to the map, he saw that Thynia inscribed one side of an immense triangle, bordered on the other two sides by Sidon Flexus and Delphi Flexus. The curious arcurate forms of the great flexus fissures came as the price of Europa's migrating crust. The ice surface was not locked to the rocky core, but rather floated freely upon a deep ocean. With each circuit of Jupiter, the moon's ice shell swiveled and shifted over the liquid. As fractures began to form, the combination of the moon's rotation, its slightly elliptical orbit, and the movement of the ice crust caused the fractures to crack in a slowly turning arc, leaving an elegant calligraphy of fine lines drawn onto the surface of the bright moon. At the point of two of those arcs along Sidon Flexus, deep within the high radiation region, lay one of the most likely sites of geyser activity. Along the cycloidal fracture spread a series of overlapping rings of fresh, powdery ice, the kind that rimmed the geyser-infested 'tiger stripe' canyons of Enceladus.

Orri's ground-penetrating radar chart detailed the subsurface ice within the triangle. It was complex. Over the years, some theories had held that Europa's crust was quite thin, a skin of frozen water a few kilometers thick, floating over a 100-kilometer deep ocean. Those advocates saw the chaos regions, dimples and hollows as collapse features over deep water, like sea ice in the arctic. Other models suggested that the crust was thick, with slow-moving diapirs—blobs of solid ice—migrating through the crust and creating ponds within the ice but cut off from the oceans below. Naturally, both camps were partially right. Europa's labyrinthine underground was a warren of chambers and water-filled shafts. Some chaos regions did, indeed, rest atop isolated underground ponds, but others represented windows into the abyss. He was betting that a small chaos region next to Thynia Linea was one such site. A lot was riding on it. He'd been the one to suggest it, over objections from several other experts in the field. If he was wrong, it would be an expensive mistake for the team and a blight on his record. He tried to push those thoughts away.

He found the most recent data encouraging. They revealed a series of chambers leading from the surface to the liquid sea just 8 kilometers below. The route looked like a pinball machine. But once in the water of the upper chamber, the descent appeared to be smooth sailing. Unfortunately, the map was five weeks old, and some of the chaos regions had proven to be quite mobile. It was possible that Europa's shifting plates had completely cut off the route by now. The next orbital radar pass was scheduled just days before the team left for the south dome. If the route was impossible to navigate, they would need to look at other possibilities in the region, but Thynia Chaos was clearly the best option as things stood now.

He jotted down some notes. There was so much to be done before departure. He hoped Felicia Tanaka would come along. As he finished his little survey, he forwarded the data to Tanaka under the heading, "Just in case you decide to join us."

Gibson sat back, arched his spine and rubbed his weary eyes. He reached over, took a sip of cool coffee. His thoughts turned from today's ice maps to the frozen wastelands a decade ago. Something nagged at the base of his neck, and wouldn't let go. He scrolled through the history of the lost expedition, and then turned to examine the recent orbital shots of the abandoned minirover. It sat on a ragged slope some 250 kilometers due east. Even for a slow surface rover, it lay less than a day's drive from the original dive site. Little was left of the dive outpost itself. The dome, originally at the edge of the Adonis Linea triple band, was long gone. Search teams had recovered several bodies there, and others on the rover that had made it farther north. But what was this little rover doing so far away from the normal route? Had they gotten lost? Suffered deep space narcosis en masse? Were they heading away from the hot zone, fleeing the radiation?

"Why?" he murmured. "Why there?" He pulled back on the orbital image. The short-range rover had left evidence of its route: broken ice, cleared pathways through boulder fields, even tracks in some places where the frost lay. Tracks, still there after more than a decade. Whoever was driving had made an intentional bee-line straight across the landscape. To what? What was so special that they would go out of their way like that?

He expanded the image, and began to search just east of the rover. There, the terrain rose in a series of parallel rounded ridges, flattening out at their apex into a triangular plateau. Faint lightning bolts of fractures raced across the plain, some white, others a rusted brown. Unlike the much larger triple bands or linea, all of these seemed to be painted on to the flat landscape there, a giant artist's snowy abstract. In the image, boulders and ice domes cast long shadows in the low sunlight. One shadow in particular seemed out of place—more irregular, more lopsided. Gibson zoomed in. At the front end of the shadow, amidst vast tracts of Prussian blues and neutral grays, gleamed a bright orange spot. He zoomed in closer. The resolution of the orbiter's imaging system was less than half a meter. Gibson could just make out several round orange and black objects, each a few meters across. At the center, linking them together, was a shiny object, a hub of some kind. The assemblage was too small to be a shelter or habitat, but it was large enough to be a supply cache. Supplies. Emergency supplies of food or oxygen or electronics. Perhaps that's what they were after. But why stash the provisions out in the middle of nowhere, far beyond the scheduled route, in the first place?

He ran his fingers through his hair. Perhaps he would have to be satisfied with yet another mystery that would never be solved, cloaked behind the barricades of time.

V

The "urgent" chime awoke Hadley from a badly needed nap. With Europa's 42-hour nights, it was easy to nap. The monitor ID read Tefia Santana. Hadley hit the reply key. Santana's face appeared on the monitor.

"Sorry to bother you, but I thought you'd want to know ASAP: Ted Taaroserro came in to the medlab a few minutes ago for an infected cut."

"That sounds fairly pressing," she grumbled, brushing the hair from her eyes.

"Hadley, he has cuts all over the backs of his hands."

Suddenly, Hadley was wide awake. "Cuts? More than one?"

"Dozens. On his forearms, too. I've seen it before."

"Are you thinking DSN?"

"Deep space narcosis affects less than 3 percent, but if the situation is stressful, the numbers go up. I'd say you guys have had a bit of stress lately, wouldn't you?"

"In spades."

"I'm telling you, if that's what this is—if these wounds are self-inflicted—you need to be cautious. Watch him. If he goes over the edge while you guys are down there, away from Taliesin—"

"It could be a problem."

Tefia leaned toward the screen. "It could be deadly. For all of you."

Hadley clicked off. She felt the gathering of storm clouds, the growing intensity of claustrophobic panic. Beyond that was the sorrow at the loss of teammates and colleagues. Her hands ached. She noticed she had balled her hands into fists. She tried to take in a deep breath, to relax her fingers, but the tension wouldn't go. She grabbed her pillow and threw it across the room.

6

Dark Curtain

I

As Taliesin approached local midnight, Jupiter's full disk stood ready to begin its periodic retreat, withering away from gibbous toward a thin fingernail. Hadley wondered if there were any creatures beneath the ice akin to the Earth's mythical werewolves. If so, they would be baying tonight. The blackness outside seemed to be filtering into her mood. With two days to go before departure, Hadley had been checking and rechecking equipment on the rovers in the hangar deck and on the sub drones in the lab. She had tasked Ted with looking over the human-rated submersible, and he had done so, he said. He seemed stable, but he kept his sleeves rolled down.

She headed back toward her room, still feeling unsure of herself. All that equipment constituted so many variables, so much complexity, so much to go wrong. And those problems of hardware only added to the personnel issues. The habitat's environment fans sang a melancholy sonata, sounding like distant voices screaming and chuckling.

A wad of paper fell to the floor as Hadley opened her door. She reached down and picked it up, glancing down the hall. No one was around. Perhaps it had been there a while. She unfolded it. On the outside someone had scrawled the words, "In light of recent events." That someone was Ted Taaroserro. She recognized his delicate handwriting.

She unfolded the note carefully and squinted at the faint lines.

You asked about Amanda and our dinnertime conversation. Well, it wasn't just Amanda. I decided that it is better if I tell you about it before we embark on our

© Springer International Publishing Switzerland 2017
M. Carroll, *Europa's Lost Expedition*, Science and Fiction,
DOI 10.1007/978-3-319-43159-8_6

little field trip to the south dome. I was not sure of myself until that night. It only made sense to me when I did some digging. Can we talk tomorrow? Perhaps somewhere public but away from the crowds? No electronics? T.T.

The note raised the hair on the back of Hadley's neck. It sounded a lot like Joel Snelling's whisperings to her. But Ted's note had been old school, written on paper and crammed through the small gap in her door. After his initials, he had jotted down a P.S.:

We owe it to Joel and Orri.

She stared at the piece of crumpled paper. What had he seen? What had she missed? Again, she thought back to the trip out, and the various mealtime conversations. Amanda had not been the only one who was startled at Ted's forgiving spirit toward the darker participants of The War. And another question bothered her—how much of this was pre-DSN paranoia? Perhaps Ted had spoken with Orri. Hadley was up to here with the cloak and dagger stuff. Departure to Thynia Linea lay just a day away. She picked up her headset and dialed Ted in.

"What do you want?" he snapped.

She tried to sound a note of calm. "You're the one who wants something, right?"

"Look, Hadley, this place is monitored," his tone softened. "We have to meet."

"Ted, I'm tired of waiting for things to happen. Please tell me what's going on."

A long pause met Hadley's earpiece. Finally, Ted whispered, "Something's been bothering me about one of our dinner-time conversations on board ship. Did you see how they were looking at each other? Something when we were talking about The War, after Orri mentioned the camps?"

"Not so much. We who?"

"…It was subtle, I grant you, but I need to bounce it off someone who is reasonable. You are the one who comes to mind. Can we meet? Just the two of us, please?"

He sounded plaintive, almost pathetic. He was running scared.

"Where?" she finally said.

"They usually abandon the lab around this time of night. If the lights are out, we can chat just inside the door."

"Okay. Now?"

"In an hour. Good enough?"

"Good enough." She disconnected, and reflected upon how she had *not* warned him to be careful, to watch out for himself. It dawned on her that she didn't trust him. Was she becoming paranoid? Perhaps, or maybe she was simply being prudent.

–*–

"I thought you'd never ask." Gibson stood at the window in Hadley's room, gazing at her with furrowed brow. "No more of these intimate meetings. It's best for everyone's sake."

"I completely agree," Hadley said. "That's why I need you to follow me. At a discreet distance, of course."

"The way Ted's been distracted lately, I'm not sure he'd notice if I gave you a piggyback ride into the room. He's becoming as sullen as Aaron."

"Yeah, he does seem awfully glum. Tefia suspects he's in early stage deep space narcosis."

"Well that's just great. As if we haven't had enough trouble on this thing."

Hadley gaped at him, then grinned. "Hey, what happened to my cheerleading unit?"

"He just went south for the winter. Ted's a can-do guy with an incredible CV. Great researcher, in-demand world lecturer. Man of faith, always encouraging people. Now this? I've always thought of Ted as a sort of moral compass for the group. I guess everyone has a dark side."

"Which makes it all the more important for you to come along."

"I never took Ted for a murder suspect. I guess it wouldn't be the first time a wolf dressed in sheep's clothing."

"In this case, shepherd's clothing. Look, I'm supposed to be there in five. Why don't you wait for me here for a couple minutes and then come listen at the door. I'll try to keep us near it."

"Deal."

She stepped out of the door, into the dark and very long corridor, and made her way toward the lab. The overhead lights had changed to the red of the night shift. At the junction of two habitats, she tried to remember which way to go. All the hallways looked the same, prefab and a little ragged around the edges. She just couldn't remember. She took the left hallway.

As she made her way down, she could tell she was going the wrong direction. Storage lockers lined sections of wall, and many of the doorways had the rounded corners of spacecraft hatches rather than living quarters. Perhaps these opened into the backside of the place she needed to be. She unsealed one at the end of a short passageway. Rather than the far side of the lab, she had stumbled upon something remarkable: a vast warehouse, a junkyard with parts of wrecked rovers, sections of partially cannibalized cranes, and battered storage crates. With all the vintage equipment, the place could have been a museum. On its far side was a garage-like airlock that led outside, marked by the standard CAUTION: VACUUM ACCESS labels. The chamber was frigid. Years of dust had accumulated on every flat surface, bearing testament to the generally stagnant airflow of a storage area. This was clearly the wrong place. She turned to retrace her steps.

_*_

Gibson had started to wonder if the chronometer was running slow. Five minutes seemed like a long time, and Hadley's room became awfully stuffy. A lot could happen in five minutes. He looked at the clock for the fourteenth time. "Close enough."

He turned toward the door when the chime went off on Hadley's monitor. "Hadley, you there?" He saw the concerned face of Tefia Santerro.

Timing was everything. He couldn't believe it. He tapped the screen. "Hi, Tefia. She's, ah, out right now. Can I help?"

"As a matter of fact, I think you can. I called because Hadley's room is closest to the medlab and I have a little problem that requires three hands. I, oh!"

A flash of light momentarily whited out Tefia's image. As she faded back into view, she was holding a piece of equipment Gibson didn't recognize. He did recognize the look of sheer panic on her face.

"Be right there!" He lunged through the door, feeling torn between the crises on both ends of the hallway. But what could he do? Tefia's emergency was more than potential. He bolted into the medlab. Tefia stretched across a table, desperately trying to close a valve. Flames poured up from beneath a shelf next to her, licking at her legs. Somebody had apparently disabled the fire suppression system in this part of the lab, probably because of the delicate equipment, Gibson guessed.

"Nothing is supposed to burn in here," she hollered.

Gibson grabbed a manual extinguisher and aimed. It sensed what combustible material was active and automatically adjusted a spray across the flame, which disappeared instantly.

Tefia let out a long breath, and wiped some strands of wayward hair from her forehead. "That was just a little too exciting."

"What was it?" Gibson said.

"It's happened in other labs, when you do certain testing and don't follow certain procedure. I think it was my medic. He's new."

"I'd say it's time for a training session. Are you okay?"

"I'm fine. Thanks, Gibson."

"You're more than welcome."

A young man plunged through the door, out of breath. "I came as soon as I heard the alarm. What happened?"

Tefia smiled wanly. "Gibson van Clive, meet Benjamin Reynolds, my Med Tech. Ben, this is Gibson, who just helped me save us from a fire that could have been a whole lot worse."

Ben shoved his hand toward Gibson, who took it.

Tefia's hands settled on her hips. "As for what happened, someone forgot to follow some important protocols."

Ben blushed.

"I'll let you two confer," Gibson said quickly. "Dr. Santana, if you're sure you're okay, there's someplace I need to be."

As soon as she thanked him again, he dashed down the hallway to the lab. The door stood ajar, light spilling from the interior. Inside, he heard only deathly silence.

"Had?"

Hadley peered around the edge of the door. "Where the heck were you?"

"Putting out a fire. Literally. Is he ... are you ... ?"

"I'm fine. Ted never showed, and I'm nervous. He's not answering his wristcom. We need to find him. Now."

"He probably shut it off for some peace and quiet. I've been known to do that."

Hadley frowned.

Gibson said, "His room, maybe?"

—*—

"Ted, you there?" Hadley rapped on the door after hitting the chime several times.

"With these walls, he can hear us. He's not there or he's not answering."

"Open," Gibson commanded the door. It was unlocked, and slid aside.

Dim light filled Ted's quarters with a disturbing solitude. The man had a sparse room, decorated with a handful of scrolling photo albums showing people and places from his illustrious travels. On his desk, the monitor displayed graphs of radiation.

"He's been doing his homework," Gibson said. "Maybe he went to your room."

Hadley turned and made her way quickly down the corridor. She couldn't shake the feeling that time was of the essence. Her door slid open to an empty room.

She scowled at Gibson. "Where in the worlds could he be?"

"It's a pretty big place. Lots of nooks and crannies. We can't send out search parties, can we?"

"I feel … responsible. If anything happens to him after—"

"Hey, hey," Gibson interrupted, taking Hadley gently by the shoulders. "You are not responsible for Joel or Orri. Or the thing a decade ago. I keep telling you."

"My watch," she said, looking down.

"Nonsense. These events were outside of the purview of our expedition. As for Ted, he's a grownup, and he's careful. Let's not panic."

"Gibs, I want everything to go right, and so far … "

"Everything will go right. It will be a spectacular journey of discovery. Lewis and Clark. Armstrong and Aldrin."

"Barnum and Bailey," she added, smiling up at him. "It's been a circus. You're right, of course. I just wish we knew where Ted was."

"It's late. He'll undoubtedly check in first thing in the morning."

"And then I'll kill him," she grumbled.

"I'll help. But in the meanwhile, I want you to promise me something."

"Oh, yeah?" she said skeptically.

"Yes. I think from now on we should stick together, like I said. Don't split up, even for a minute. No more of these late-night one-on-one meetings. Okay?"

"Probably a good idea. I'm going to bed, but I'm keeping my com open in case he checks in. Breakfast at seven?"

"See you then."

II

It was dark outside, part of the 40-some-hour-long Europa night. But human time marched to its own drum, with only the red lights of the habs to provide a clue as to when night was to be celebrated. Hadley had a difficult time convincing herself that it was morning, or breakfast time. She pinged Ted's room again, but there was no answer. She pinged Gibson.

"Up for one of those Denver omelets they serve?" She tried to sound cheery.

"You bet. Meet you in the hall."

She stepped into the corridor a few doors down from Gibson's quarters. He had just come out. As soon as they were close, Gibson lowered his voice and said, "Any sign of the good Dr. Taaroserro?"

"Not a peep."

The two proceeded down the corridor in worried silence. They heard the cacophony of mealtime long before they made it to the galley door. The place was packed. The dense air smelled of cinnamon, eggs, burnt toast and coffee. As they jockeyed their way through the door, Gibson asked a departing tech, "Hey, what's all the commotion?"

"This group is going to the west today on some research outing or other. They're all having breakfast together. I just work here." He shrugged to emphasize the point.

Gibson turned to Hadley. "Should we try the other galley?"

"It's a glorified snack bar."

"Yeah, with less of a line."

"Maybe so." As she turned back toward the door, she spotted Ted. He was extricating himself from the mob at the front, taking a tray with him and heading for the other door.

"Ted!" she bellowed.

He glanced over his shoulder and spotted her, offering his trademark disarming smile.

"Oh, Hadley," he called over several heads. "I am so sorry. I got held up. We'll chat soon." He disappeared out the door. Half of Taliesin's population seemed to be wedged between Hadley and the exit. Ted was gone.

She looked at Gibson, incredulous. "Chat?"

"Does seem a bit cavalier," Gibson said.

"We'll chat all right. I hope he's not trying to back out. After I get some calories in me I'm going to his room."

"I'll just stay back and eat my breakfast, let the two of you set off whatever fireworks you need to. Except that will break our new rule."

"I think, considering it's Ted, we can assume the best. But how about if I keep my wrist monitor on speaker and you can listen in."

"That will work. I feel kinda guilty eavesdropping."

"It's prudent." She tapped her wrist. "Besides, this thing is so old you probably won't hear much."

"Had, he looks bad."

"Yeah? I guess I was too furious to notice."

"I don't think he's doing well. Did you see those bags under his eyes? Looks like he hasn't slept in days."

Hadley softened. "Poor guy. Maybe he really is getting the shakes. We'll need to get him some help."

"You guys on that expedition to the south dome?" The speaker was a bulldog of a man Hadley had not seen before. He sat at a table, twisting in his chair to face them.

"That's us," Gibson said.

"Heard about the death of your team member, Joel Snelling."

Hadley braced herself for another condolence speech, the kind she had endured a dozen times in the last day. But the man hissed.

"Couldn't have happened to a more deserving subhuman. That guy let a few of my best friends die in that dome collapse back a few years. He should have been banished from this place."

The man turned his back on them. They left him to eat by himself. As they exited the room, Hadley leaned over and whispered, "Looks like Joel had at least one enemy right here."

III

Ted had left the lights off in his room. He lay in bed, his tray of food next to him, scarcely touched. He had a washcloth over his eyes. He seemed restless, but he didn't get up.

"Do you think you'd feel better if you ate something?" Hadley tried not to sound too motherly.

Ted merely groaned. He turned his head toward the window and peeked out from under the cloth. After the long night, Jupiter cast its warm glow upon the icescape outside. "Out of the north he comes in golden splendor and awesome majesty."

"He?"

"It's from the book of Job." He rolled onto his back and faced the ceiling again. "Very ancient book, did you know? Probably written before any of the rest of the Torah was. Before Genesis or the Psalms or Proverbs."

"Ted, where were you? Where did you go last night?"

"Very old," he said. Had he even heard her?

"Ted, we were supposed to meet. Where were you?"

"I was scared. I think they're on to me. Didn't want you to get in trouble, but I couldn't make my stinking intercom talk with your room."

He slurred his speech. His were the classic symptoms of deep space narcosis.

"I'm going to get you some help, okay?" She tapped his screen, then hit the comms button manually. Nothing.

He took in a breath. "I do not lose heart. Even though the outward man is perishing, yet the inward man is being renewed day by day."

"Ted, focus. You're right. There's something wrong with your room's comms. I will be right back."

"You know what my real name is?" Hadley froze in the doorway. "It's Isaac. It means 'he laughs'. I guess I came out of my mom that way. Laughing. That's me. Always laughing at life's celebration." He paused, then added weakly, "I just don't feel much like laughing lately."

Hadley stepped back to the side of his bed. "What can I get you?" She pulled the cloth from his eyes and used it to wipe his forehead. His clammy skin misted with sweat.

"I feel like a dark curtain is drawing all around me. It's hard to see out sometimes. I know I'm sick. I need healing. Jehovah Rophe. God is my healer. Tefia is my healer." His eyes became clear, a look of distress on his face. He grabbed Hadley's sleeve. "Together for strength ... together for power."

She knew the phrase, of course. Something from history—a famous slogan. It was something from the past, but she had heard it again more recently. She gently removed Ted's fingers from her shirtsleeve, and grazed his forehead with her thumb.

"Hadley, want to see a secret?" he rasped.

"Is now the time?"

"You'll like it." He moved his arm, limp-wristed, toward the small bureau in the corner. He still had the bandages across the backs of his hands; self-inflicted wounds. DSN. "Second drawer down. I want you ... to see." His breathing became labored, but from physical exertion or psychological stress, Hadley couldn't tell.

She reached over and pulled open the drawer. A worn, leather-bound book lay inside. She reached in and opened it. She was silent for a moment as she processed what she was seeing. "I had no idea."

"I don't make a habit of showing off. Just a hobby of mine."

"Sure, but there's such a thing as hiding your light under a barrel."

"Bushel," he corrected.

Across the opening page, a delicately drawn landscape danced in fine lines and swirls. It was a landscape she had seen, one of the most famous overlooks on Mars, at the edge of Melas Chasma. The light tumbled softly down the cliffs, catching edges of boulders and spines of ridges. Gullies drifted down into the shadows, gentle blizzards of graphite. Page after page marched before her: Hawaii's N'a Pali coast, the rim of Stickney Crater on Phobos. She held up a sketch. "Ganymede?"

Ted blinked in the darkness. "Yessss. On my last trip out here. Harpagia Sulcus. God's fingerprint."

"They're beautiful."

"I want to sketch our first sight of the geysers. Mm."

As she turned the page, another sketch spread before her, but this one lacked the delicate line, the loving caress, of the others. Harsh pen strokes cut across the page, detailing a panorama of ruin and smoke. Within the smoke, drifting across the sky, smiled the rictus of a skull.

"Oh, that." Ted said from his bed. His head was turned. He looked at Hadley. "The War. It still haunts me sometimes."

"I understand."

"Yes, I think you do."

Hadley tipped her head to one side, studying him. "What do you mean?"

"I think you had more dealings with The War than you shared with our little group. I think you have some deep hurt there."

Suddenly, without warning, the tears came. Her cheeks warmed as her chest tightened. "You're pretty good, Ted."

He said nothing but waited. The ball was clearly in her court. She decided to humor him.

Hadley drew in a long breath. "You know, my parents were remarkable people. While everyone was watching those political winds tossing national borders back and forth, while the bombs fell and our computer networks and power grids disintegrated, they were there, together. Lighthouse in the storm."

"Where were you when it all broke out?"

"Doing a geology stint on Ganymede, stratification of the graben along the rim of Asgaard Basin. I got called back home when the Sudamerican Federation fell. And then they sent my father away to … to do whatever an engineer does in a war. And then he was lost. Just like that."

His eyes glistened in the darkness. He nodded, just slightly, as if to encourage her.

"So after a year of silence, we began to hear rumors that he had been taken to a camp. When The War ended, we got the news we'd all feared. He had died at one of the camps of the Seven Sisters. I prayed he hadn't become one of their experiments. The thought nearly drove my Mom mad. Daddy. Steven Nobile. From a long, illustrious line."

"I am so sorry," Ted sighed. "So much darkness in the world. At times." He nodded toward the porthole. "Look out there."

She did. From Ted's window, she could see a low-lying ridge with dramatic scarring along the cliff face. Ramparts of ice reared up behind it, like a parade of dominoes splayed out in grays and blues.

Ted's voice took on a strength that hadn't been there before. "You see, some scientists, like Gibson, for example, would look out there and see ices forced up by complex tectonics, symptoms of the gravitational … war, I suppose, yes, battle … going on between Jupiter and the other moons. And that is all they would see. Beautiful tectonics. Forces of nature. But what if there is more? Something deeper? Something … spiritual? A design behind."

"A Creator behind the creation?" Hadley offered.

Ted nodded, gazing at the ceiling. His eyelids drooped.

A smile flickered across Hadley's lips. "I wouldn't sell Gibson too short in the spiritual department." She frowned and spoke gently. "But in the end, what's the real difference? They both result in the same outcome—an appreciation for the natural wonders."

His eyes closed and he smiled. "Ah, but do they? You see, you can run the tape back as far as you want, to Europa's evolution, to Jupiter's accretion—" The clarity returned to his voice as his eyes opened again and studied her. "Back to the beginning of the Solar System in a shock wave from a nearby star or supernova or what have you. But eventually, we are all faced with first cause. What began it all? And you can go the 'multiple universes' route." He said it disdainfully. "But it's just like Francis Crick saying life came to Earth from outside in an episode of panspermia. You haven't solved anything, just moved the timeline back. No, Hadley, if there was a trigger behind the First Cause, something intentional at the beginning of all things, then that becomes something more than academic."

Intrigued, she leaned closer. "How so?"

He coughed. "A moral component. An aspect that says, 'Here, then, is a better way to live, a way to freedom.'" Ted rocked his head back and forth, his words garbled, feeble again. "But God is just, too. Will he forgive me? Has he? Of course he does, through his Son. But will they?" He flinched. "It's the voice that gives it away."

"What?"

"I could hear it. In their voices. Like the song...of the meadowlark. Each one. Different. Look, Hadley. Listen."

"Listen," she repeated.

He grabbed her sleeve as she turned. "Breakfast. It wasn't dinner. It was at breakfast."

"Get some rest. I'm getting Dr. Santana."

He nodded and closed his eyes. As Hadley reached the door, she heard him moan, "If I go up to the heavens, you are there; if I make my bed in the depths, you are there. Hadley, I deal with hurt from the War, too." And then he told her something else.

—*—

Tefia leaned in close to Hadley. They watched Ted through a glass pane, but whispering seemed appropriate in his delicate state. "I've given him some new medications to stabilize him. It's a sort of pharmaceutical cocktail they approved fairly recently. They say it's pretty effective."

"Why didn't we give him all that stuff at the first sign?"

"DSN symptoms can be misleading, and the complex of meds is pretty hard on the liver. I had to get his permission. He felt it would be a good risk, and I think he made a good choice."

Hadley fixed her with a stare. "But is he safe? Can he be trusted?"

"How badly do you need him?"

"He's a critical part of the mission. He knows the radiation fields as well as you do. He's an expert submersible pilot. Without him ... "

"All right. If he remains stable through tomorrow, I would say yes, definitely. He should improve steadily, and I'll bet he'll be able to work. We'll watch his blood levels, and he's under 24/7 observation on vid. If his blood chem plateaus, I can declare him good to go. I just wish he had a couple more days."

"Me, too," Hadley said. "But the geysers of Europa won't wait."

7

Departure

I

A glider was an impressive sight. Twenty meters from stem to stern, loaded with the supplies of an extensive expedition, and carrying small all-terrain vehicles in its bed, the heavily laden craft still seemed sleek and capable. Two gliders in a row provided a downright spectacle.

Standing at the observation deck, gazing through a panorama of bay windows overlooking the staging area outside, Sterling Ewing-Rhys shook his head in admiration. "They sure know how to put on a good road trip."

"That they do," Dakota said. She stood on one foot, a martial arts habit that often looked like ballet, and sometimes made those around her uncomfortable. Sterling seemed to take it in stride. Here, rovers came and went, shuttles arrived from incoming flights, and general bedlam seemed to ensue, day and night.

"It won't be long now," she said. "Are you excited?"

"I've never been a fan of packing."

"I mean about getting there. About running all your little robots." She wiggled her fingers like a piano player.

"One of those 'little robots' is 10 meters long and will have a couple people in it! I can completely control our submersible from the comfort of my console." He rubbed his hands together, hunched over, and let out an evil chuckle.

"Your nefarious nature is showing through again. I'll keep it in mind when it's my turn to go down. If I get a turn."

© Springer International Publishing Switzerland 2017
M. Carroll, *Europa's Lost Expedition*, Science and Fiction,
DOI 10.1007/978-3-319-43159-8_7

"I'm sure you will, but you can request sampling at any time, and we can do it from our little control center while the people on board are going about other business."

"So while Hadley's diving for her soggy volcanoes, I can be taking water samples remotely?"

"From our console, yes. Or catching your critters, if you want to be a real optimist. As long as it doesn't preclude the passengers' operations. It's like we have two teams, one piloting in person and another piggybacking, along for the ride robotically."

"You sound like you don't need any people in the mix."

Sterling sniffed. "It's tempting. Sitting in a radiation-protected dome is a lot less exciting than actually diving through miles of ice water, and that's the kind of excitement I can do without."

"Yes, but when I get down there, I know what I'm looking for. I have reflexes, flexibility, intuition."

"A drone operator has intuition, too," Sterling countered.

"Sure, you have human involvement either way, but there's nothing like a person on site. You're just jealous," she teased.

"I'm sure that's it, *Miss Barnes*."

Outside, floating less than a meter above the ground, the close-in glider inched forward, revealing its twin behind. "There it is!" Dakota pointed to the far vehicle. On the back of its flatbed, the bright red submersible hung from a crane, secured by cables to the glider's platform.

"Reminds me of a red chili pepper. We're fond of those down south."

Dakota looked at the antenna-like sensors projecting from the nose. "I was thinking something more along the lines of catfish. I've had those, deep fried. Very tasty."

"It may be a southern dish, but I'm about as much a fan of catfish as I am of packing."

"What kind of a southern boy are you?"

"Gentrified," he said, letting out a long breath. "I wonder how Hadley's doing with the prep."

"And everybody else. Have you heard how Ted's doing?"

Sterling rubbed the back of his neck. "He's back in his room, so he must be doing better. Dr. Santana said he seemed to be 'responding well to his medication regime,' whatever that means. We should check in on him, see if he needs anything. Tomorrow's the big day, and I'll bet he has some catching up to do."

Sterling and Dakota made their way past the control center and medlab, on to the personal quarters section. The area had a distinct smell, of plastic and metal, of dampness, and a slight whiff of locker room.

Ted's door was closed. Dakota knocked. "Hey, big guy. Can we come in?"

After a silence, Sterling said, "Maybe he's gone out. That's a good sign."

"Yeah, it is." Dakota's wrist comm pinged. She bent her elbow, her arm chest-high, and said, "Barnes."

"Dakota? Hadley here."

"Hey!" she said with as much unbridled enthusiasm as she did everything.

"Can you come look over the manifest? I want to make sure there's nothing missing down at the dome that we'll need to bring with us. There probably won't be time for them to send a drop ship or rover down once we're there."

"Roger that."

—*—

With Europa's arctic vista out the window, and the crew loading cargo aboard the gliders, Dakota sat at a counter with Hadley, sharing a small monitor. Hadley jabbed a finger at the screen. "I've got to hand it to the Taliesin gang. They've been efficient. Look at all this equipment they've got in place down there."

Dakota clicked down the list. "Biome energy flow, organics, robotic sampling devices, miles of cable and antennae for dropping down the shaft... GCMS... yep, it's amazing."

"See anything that's missing?"

Dakota leaned in, as if she could read between the lines. She shook her head. "They've already got a lot of stuff I thought we would have to bring."

"Good for them," Hadley said. "We may need to do some organizing once we get down there. The robotic construction isn't completely competent, but they've done a great job. Most of it was remote."

"But they sent one crew down, right?"

"Yes, just a month before we arrived. They put on the finishing touches, checked the bore hole and such, but they weren't all scientists. I'm sure everyone on our team will have a few adjustments they want to make."

"Fine with me," Dakota said, leaning back from the screen. "Missing equipment would be tough. Adjustments I can handle."

"I think we're in good shape. Still need to have Sterling go through the manifest, and get Gibson's final sign-off."

"Yes, of course." Dakota said it in a subdued tone. Hadley noticed but didn't comment. She had the strong suspicion that Dakota had a thing for Gibson. She tried to envision them together. For some reason, she had a hard time picturing Gibson with anyone. Since the loss of his wife, he seemed a confirmed bachelor.

"Ted's all squared away?" Dakota asked.

"I think so. At the dome, his main focus will be submersible and ROV operations, but they left us with all sorts of radiation gear down there, and he'll be monitoring the various remote stations and our home field generator."

"Too bad you can't bring that thing with you into the hot zone."

Hadley nodded. "Yeah, but we've got good suits, and we'll set up a little storm shelter using the glider's cab and ice bricks. They make great radiation barriers. We'll be fine, and it's only for a couple days. I'm sure you'll miss us, just you and the Grants and Sterling in that little dome. And Jeff Cann's staying behind, too."

"I'm pining away already. Where is Ted, by the way?"

"In his room. Doctor's orders."

Dakota leaned forward conspiratorially. "I think he's being a bad boy. Sterling and I stopped by and there was no answer."

Panic flickered across Hadley's face. She keyed her monitor. "Ted, are you staying quiet, like a good little patient?"

Silence.

"Ted?" Hadley glanced at Dakota, and then out the window toward the frenetic activity.

"It's probably nothing," Dakota said. She stood up. Hadley noticed the athleticism of Dakota's body; every muscle tensed. "Should we, ah—"

"Check?" Hadley said, standing so abruptly that one foot momentarily left the floor in the 1/8 g. "Maybe so."

The two bounce-walked briskly through the warren of corridors, each not wanting to rush too much. Hadley hit her wrist comm. "Tefia, are you there?"

"Right here, Hadley. It's not as if I have a life in this place."

"Is Ted there with you?"

"No, I discharged him to his room this morning. I told him to stay put. Have you checked?"

"On our way now."

"Everything okay?"

Hadley controlled her breathing. "I'm sure it is." She signed off. At the door, she reached for the doorbell, but Dakota rapped her knuckles on the hatch and called out.

"Hey Ted. You in there?"

They waited for ten seconds. Twenty. Hadley punched the button and the door began to move, but it jammed on something.

Dakota covered her mouth with her hands. "No!"

Ted Taaroserro's foot wedged against the door. His body sprawled into his room, the lights still off.

"Lights," Hadley hollered. The room brightened. She grabbed the edge of the door, shoved her shoulder against it and forced it open.

Dakota leaned over him and knelt down. "Ted, buddy? Wake up." She patted his cheek. Hadley pinged the medlab on the emergency channel.

Dakota rolled Ted over. She felt the side of his neck. "No pulse." She shoved him flat onto his back and began to do compressions on his chest. It was primitive, but would have to do until the med techs arrived. Hadley grabbed the back of his neck and tipped his head up, preparing to give him mouth-to-mouth resuscitation. But something held her back.

"What are you waiting for?" Dakota wheezed. Suddenly, she saw what Hadley had seen. She stopped pushing. "What the . . . "

"Like Orri, only worse."

"Those sores. The blisters. Don't touch him."

"I doubt it's catching," Hadley said. But she knew that this deadly sickness was contagious—terminal, in fact—to a very select group, her group, destined for the isolated southern pole of Europa.

–*–

"He was so . . . " Dakota sniffled. Sterling patted her gently on the shoulder. It was the closest his conservative upbringing would comfortably allow.

"Kind? Wise? Tall? Yes, he was. I liked Ted very much."

Dakota wiped her nose. "Me too. I just wonder what was wrong with him. He looked like he had something. Like he was sick." She looked at Sterling. "Didn't you say you were a medic in The War?"

A bitterness fouled Sterling's laugh. "They gave me a glorified first-aid class. I was ill-prepared, to put it mildly. But I suppose nothing prepares you for all that. At any rate, if Ted had something, I couldn't tell."

Dakota looked away from the body. "Poor Ted." Her gaze focused on something unseen, something far away.

Sterling had no words to offer. The loss of Ted seemed to have paralyzed his brain functions. He tried to direct Dakota back to a less emotional subject.

"Where were you during The War?"

"Here and there." She acted as though she didn't want to talk about it, but then she refocused. It was something to ease the moment, substituting one pain for another, more remote. "I kept thinking that the news would get better. I couldn't believe the Eastern Alliance would go any farther than the African continent. I thought they'd stop there." Dakota's tone began to ramp up. "Then there was the bombing of the Smithsonian and the Louvre and the razing of the British Museum and all the rest, all the treasures and the bits of civilization, gone. And then the Western group had their own little reign of terror—rain of terror, I suppose—across Europe. What a mess. And what a surprise. Just when you think people are coming to their senses."

"There's no sense to war. Not like that."

Dakota shook her head. Obviously, they were trying to find comfort in conversation, no matter the subject. "Where did you end up? After?"

"Spent a year in AustraloNesia, then on to Mars for the last few years."

"Mars? What were you doing there?"

He dismissed her question with a wave. "Boring stuff."

"You had lots of company. That's where the Seven Sisters ended up, isn't it?"

"They caught five of them there. A modern-day Nazi Brazil, I guess. Last I heard, two were still free. Somewhere."

"Bet they don't feel free," Dakota said, "with the entire known universe searching for them."

"I suppose with all the time gone by, the War Crimes Tribunal may lose interest."

"After all that they did, I'm not so sure."

"I suppose you're right," Sterling shuddered. "What about you? Did you have to serve somewhere during The War?"

"Thanks to my youth, I just missed the nightmare on the arctic front. But I got training. That's where I learned martial arts. Karate, Eskrima, Krav Maga. It's put me in good stead. I keep up with it."

"I can tell," Sterling said. "You look like an MMA pro."

"Thanks. I think."

"Absolutely. Meant as a compliment."

The two settled into silence, watching the hulking form of Ted Taaroserro beneath a blank, very white sheet.

II

"I've ruled out viral or bacterial causes. It's poison, all right."

"But how did he get poisoned?" Hadley demanded. Her eyes were bloodshot.

Santana motioned for Hadley to come closer. "Have a look." The doctor turned Ted's head to the side and pointed to his neck. Just under his left ear, a bright red puncture blossomed across his skin.

"Some kind of shot? Syringe or something?"

Santana nodded. "Something. Odd thing is that there are actually two wounds. One is smaller, as if the needle wouldn't go in completely."

"Tefia, this is murder."

"Clearly. But let's keep it to ourselves for now." Santana glanced around nervously. "Come back into my office," she said. The two women shuffled into the back room and sat at the desk.

Santana took in a long breath. "I'm going to check the bodies of Joel and Orri for puncture wounds, although symptoms like these often fade away with time." She leaned forward urgently. "Look, Hadley, this can't go on."

"What are we going to do? Post a guard on all of my team members until we leave?"

"That's not what I mean. As field operations coordinator, I just can't allow you out under these circumstances. You've lost too many key personnel…" Santana's tone seemed half-hearted.

Hadley experienced something akin to terror. The decade of work, years of preparation and negotiations and diplomatic shuffling between various scientific and educational institutions, hung in the balance. She looked at Santana. Santana looked back with a pleading expression, as if to say, *Talk me out of it.*

At that moment, on a remote outpost stuck on a barren ball of ice circling a lethal gas giant, both women felt the inertia of history, of something bigger than human drama. A once-in-decades event was about to take place, and only a handful of scientists were equipped to observe it for the rest of humanity. Hadley knew that if she wanted her expedition, she had to fight for it.

"True, we've lost key personnel, but we have your own crack glaciologist to replace Dr. Sigurðsson, if Felicia Tanaka goes for it, and you said yourself that Jeff, your top submarine expert, could tag along in place of Ted. All we really need now is a radiation specialist. As far as I can see, *you* are more than qualified for the job."

Tefia quieted, considering. She tapped her chin. Her eyes met Hadley's. "As for Jeff Cann, I had him practicing the sim software for your little red submarine, in case Ted didn't recover enough to go."

Hadley glanced toward the medlab. "I guess he didn't," she said sadly.

"Jeff's in there practicing right now, undoubtedly with his vat of coffee at his side. And as for Tanaka, I just gave her the official go-ahead to join your team."

"Which leaves a radiation expert."

"Which leaves a radiation expert," Santana repeated.

Hadley leaned forward. "And a damned good doctor, if anything else goes wrong."

III

Dakota stood in the doorway to Hadley's quarters, looking forlornly at her and Gibson. "Sterling seemed pretty jazzed about the whole trip," Dakota said. "He's such a typical southern gentleman, sometimes it's hard to tell. But he's in his room and I don't think he'll be coming out for a while."

"He is hard to read at times," Hadley admitted. "But Ted's loss is going to hit everybody hard."

Dakota shrugged, deflated. "I'm sure this expedition is just what Sterling needs after cooling his heels for a couple years on Mars."

Hadley jerked her head to look at Dakota. "Mars?"

"Said he was there a couple years before coming out here. From one cold place to another."

Hadley said nothing, but something was bothering her. What had Sterling's vitae said?

Dakota dug her toe into the floor. "Shouldn't we do a toast to Ted at dinner or something? It just seems so empty to go on without someone saying something."

"I think you should say a few words tonight," Hadley said. "That would be nice. Give us some semblance of closure."

"Sure," Dakota said, turning to leave. The autodoor shut behind her.

"Of course there's no time for any kind of formal memorial," Gibson said quietly.

Hadley shook her head, her hair swirling loosely around her shoulders. "Timelines are just too tight."

"We didn't have one for Joel or Orri, either."

She rubbed her face, kept her hands over her eyes. "Somehow, this feels different to me. It's like bile in the throat. I ache, deep down inside. Something feels empty where there used to be a little light before. I mean, I liked Orri, and I even had a soft spot for Joel, but this is somehow different. It seems—" she searched for the right word.

"Unfair?" Gibson offered.

"More than that. *Unjust* in the cosmic scheme of things."

"He was kind. He was forgiving, too, and not in that way that religious hypocrites can be. He seemed sincere, deep down."

Hadley's only reply was a sniff. She reached for a tissue.

"Should we do this later?" Gibson asked.

"Nope. We need to find out what we can from that diary before we follow in their footsteps. We can grieve when we're all back safe and sound."

Gibson nodded and keyed the old diary. Its text scrolled down the desk monitor. "Here's my latest untangling. The thing really is a mess. It's dated February 16, and the first portion is corrupted beyond recovery. Then these bits come."

… made a quip about Alison putting Peter "on ice." Nobody found it funny. Nothing is funny anymore, now that Peter is gone.

… lison won't make it much longer. Had to tie her hands.

"Tie her hands?" Hadley marveled. "That's getting pretty bad."

"It gets worse fast. Here's the entry from two days later."

Feb 18

We lost Serge and our submarine on the third dive. Some kind of systems failure beneath the ice. All seemed jus…ine. Telemetry good and strong. Evidence of…ant…no success in hailing him. O_2 reserves would have been exhausted three hours ago, even if he scrounged the extra O_2 from… No prospect for his surv…

"Sad," Hadley said. "But we knew this from what scant records we had."

"Yeah, but now, in the next passages, the problems are expedition wide," Gibson explained. "They're hoping for a rescue." He tapped the screen.

… was holding out hope for some kind of last-minute bailout from Taliesin's cavalry, but they knew we might have comms problems, and they don't even expect us back for two weeks. By then, we'll be lon…

"I jumped on toward the end, as you suggested. It's in pretty lousy shape, too, but I did find this:

Sobering dinner. Mendelson wrote a letter to his family, hoping someone would find it eventually. Most of it was gibberish. Donnie tried to talk us into going back to the eastern science station, but Mendelson told him he was crazy. If there is any hope for more supplies, it's at the dome, but that's farther away. Donnie tried to convince us that if we're going out, we should go out in style. He wanted us all to go back to the project. It's almost finished anyway, and someone might enjoy…

"As far as I can tell, the entry is very late, perhaps the second week of March."

Hadley lowered her voice in thought. "Toward the very end. What were they doing in the east? It shouldn't take very long to set up an automated science station."

"I'm at a loss on that one."

"So what does the timeline look like? Can you chain together events?"

"Not much new to go on, but I've gotten some insights. They arrive at the dive site in the two large rovers on twelve February, at the edge of Adonis Linea. It's been a grueling, five-day drive, but they erect the dome and another habitat right away, as we know from other records."

"Glad we'll have our gliders," Hadley interjected.

"Me, too. Radiation levels are unexpectedly high and communications unexpectedly lousy. By the sixteenth, Alison Sylven has had her breakdown, and Peter Kaminsky is dead, probably murdered or killed by Alison's bumbling of some critical operation or element.

February eighteen: they lose the submarine, with Serge Montano aboard. No more diving for dolphins on Europa. And somewhere in there, at an unknown date, one of the big rovers suffers some catastrophic failure."

"Right, the Primary 1 rover. And because they didn't start out with the third, the backup rover—"

"There's not enough room to bring everyone back to Taliesin," Gibson finished Hadley's thought.

"So at some point, they split up. Why would they do that?"

"I think they were trying to get back to that Remote Science Station B site in the east. They must have had supplies cached there. I've seen something in the orbital views that might be just that, and there's a reference to an eastern cache somewhere later in the text."

Hadley rubbed her eyes. "Maybe something to repair the rover with. But by somewhere between March 18 through 22, they were all dead."

Gibson held up a finger. "There's something else I found out about Alison's death and Primary 1, down here somewhere. It's about Station B. If I can just find it." He scrolled through the fragments of text and came to the passage.

I hate to be such a hard-assed mathematician, but with Alison gone we're down by three, and that means if Donnie can get Primary 2 to roll long distance, we might make it back to Taliesin in it. He's cannibalizing Primary 1, but there's some computer thingie that still doesn't work. Can't make it as far as Taliesin, or even close, without it. He thinks he might have one in a piece of hardware in our eastern cache. Donnie and I will try to make it out to the cache in one of the minirovers, even tho they aren't designed for that long a drive. Seems to be a theme on this trip.

"Yes. So they take Primary Rover 2 north toward Taliesin, it fails along the way, and they send a mini rover to get repair supplies at the cache. Makes sense. Keep digging."

"You got it," Gibson said cheerfully, disconnecting the diary.

The monitor chimed. Tefia's face snowed into focus.

"Hey Hadley."

"If it isn't the good doctor," she said.

Tefia's expression was all business, her voice quiet. "Why don't you trundle over to my quarters for a minute?"

"Sure thing; be right there."

Tefia's door slid open as soon as Hadley hit the bell. Tefia called her into the back room. The doctor sat cross-legged on her bed with a large, old-fashioned book, like Smaug guarding his dragon's treasures.

"A *book*?" Hadley marveled. "Way out here? What ever happened to digital?"

Tefia reached across the bed and shoved the book at Hadley. "Smell it! Feel it. Listen to the way the pages sound as they waft across the air. No electronic beeps. Just the rustle of pressed organic stuff."

Hadley weighed the tome in her hand, realizing how dear the object must be. She examined the cover. "The writings of Taliesin? I thought you'd be reviewing our route or scanning rover diagrams or something."

"A girl's got to relax, especially in the midst of all this stress."

"Yes, absolutely." Hadley squinted at the faded words on the cover. "I wondered about the name of this place."

"Taliesin Crater to the north. Taliesin Chaos to the east. One gets the impression that the namesake of this region might be important, so I figured, why not?"

"So he was a poet," she said, patting the cover. She handed it back.

Tefia nodded. "Celtic. Ancient Britain's grand bard. It's likely that he sang his songs at the courts of three British kings. Three. Can you imagine? And you'll love this: his name means 'shining brow.'"

Hadley peered out the window at the glowing landscape. "Perfect."

"I thought so." Tefia's interest in Taliesin seemed oddly placed. Perhaps she was just putting off talking about the subject at hand. But Hadley let her go on about the ancient bard and his romantic legends. Finally, Tefia dropped the book on her bedside table. She held Hadley's gaze. "Now, to business. That thing we found on Ted? As you suspected, there's another on Orri. About the same place. This one was cleaner, so I didn't see it at first. It's more efficient, as if it was administered by machine versus by hand. At any rate, that cinches it."

"Damn." She closed her eyes for a moment, and then looked back at Tefia. "What about on Joel?"

"Inconclusive. He's been dead too long for my primitive suite of instruments to tell."

"So what do we do now?" Hadley asked.

Tefia glanced from side to side, leaned toward Hadley, and lowered her voice even further. "Look for somebody with a dart gun."

—*—

The monitor hummed, its blank screen casting a soft red light into the darkened room. Hadley cradled her head in her hands. "You know, I just keep wondering who it could be. I find myself fighting off panic, but it's more than panic that's frustrating me. It's the mystery. I keep thinking there should be a pattern."

"If there is one, I can't see it." Gibson stuck out his finger and brushed lint from the screen.

Hadley sighed, her eyes focused with concentration. "Me neither. Not yet. If it's Sterling, I try to imagine what I would do next. I would not attack me, because I'm the expedition head, and I can't help but think they want us all out there, way out in the wilderness, to finish everybody off. They would probably go for Dakota. Besides you, she's the most able-bodied."

"But what if it's somebody else? Aaron, say?"

Hadley frowned. "I thought about that. But I really think Aaron and Amanda are a package deal. If one is doing it, they both are, and I just can't see it. I worry more about Amanda's safety. She's older, and Aaron doesn't quite seem the protective type."

"And besides, she reminds you of your grandmother. Not very objective."

She dropped her hands, stood, and began to pace. "Say somebody really does have a secret—an important secret. Something with really high stakes. What if Orri or somebody else put two and two together—something about the past that somebody wanted buried—and somehow let him, or her, know? What if that person was blackmailing the culprit, but his target didn't know who was doing the blackmailing?"

"All the culprit could know was that it was someone in our group." Gibson warmed to the subject. "And they might want to begin removing the threats. Systematically."

"They must have really wanted to keep their little mystery to themselves. I hate being part of a system."

"So we've narrowed the suspects down by three. Which leaves Dakota, Amanda and Aaron, and Sterling."

"And you, of course," she said theatrically. "You've always struck me as a suspicious character."

"Your parents always said the same thing."

"Are you kidding? They loved you! Mom still raves about you. Of course, she still thinks you're in high school."

"I can tell you one thing." Gibson's voice dropped, calculating and low. "If someone decided to get rid of anyone in our little party who might know something, a remote science outpost a thousand kilometers from Europa's only major base is the perfect venue, don't you think?"

"That's going to put a damper on our going-away breakfast celebration tomorrow."

"Last Supper. We have to go anyway."

Hadley stood. "Bon appetite!" She paused, frowning. "It's the past that's got my attention more than the future. I think while you're digging through your diary I'll go do some digging of my own. Somebody is not giving us the entire story here. Someone on our team is hiding something that just might come out in his—or her—dossier."

Hadley looked out the window at a landscape suddenly hostile and terrifying. "Gibs, it seems like the world becomes ripe for disaster and...and *intrigue,* when people are left to fend for themselves. When we're cut off from all the things we cocoon around us, the things that make us civilized. The things that make us human, day to day."

"It's like camping," Gibson said. "I hate camping. I'll do some more decrypting before breakfast. See you in the morning."

—*—

Hadley grew tired of her tour of the team's personal histories, but she couldn't think of any other approach. So much of the background data had been lost in the aftermath of The War. Amanda and Aaron Grant seemed to have not existed in the preceding years, and both Dakota Barnes and Sterling Ewing-Rhys had confused, incomplete backgrounds. Sterling, in particular, had made it clear that he transferred from Earth to Phobos only weeks before embarking on the expedition. And yet, he had told Dakota he had spent years on Mars. Why would he lie about such a minor thing? She wanted to know, but the sand had run from the hourglass. It was time to sleep before the big day. Departure was upon them.

IV

Gibson's eyes scanned the monitor. The text in this section was largely intact. The first entry was brief but significant.

Feb 25

Donnie and I will try to mak...the cache. Maybe it will be our salvation after all.

Feb 26

Minirover 2 gave up the ghost, but not as predicted. I was driving, and I lost it over a little ditch. Just a bump in the road, but ruined the undercarriage. We made it to the little science outpost, but nothing doing on that piece of hardware Donnie needs. They're going to try coming to get us in the crippled Primary Rover 2. Donnie says that's it. Nothing he can do to resurrect the hobbled machine. Guess there's only one person in the resurrection business, and he was a carpenter, not an engineer. We just couldn't make it far enough north in the shape it's in. It might not even make it out here to pick us up. No way we could hoof it north with current supplies, and not enough power in our secondary antenna. We're trying, but nobody can hear us in all this interference. Sweet Genevive Dupre is finally gone, after a graceful battle. We're all radiation sick. Everything seems to be falling apart.

Gibson could hear the voice of the lost explorer, hear the desperation in his words. The diary's tone became progressively more urgent.

… new lease on life. He says if we can get back over to the sub support system at the dome, we can pull out its … fit it into the rover's scrambled brain. It's a long trip, but we do have one mini that still works, so I will try to make it down there with instructions while Donnie and Mendelson work on the rover up here.

… f only we had one stinking wheel motor—just one—we might make this work. But it seems the gods are against … nie won't let me go alone. If he can … ake it down after all.

Donnie is in a bad way, what wit … I would have thought Mendelson would hold out the longest, but he's starting to get the shakes, and I give it … oesn't leave us much time. If only we had that third rover. If only we could have rigged the radio for … f only-if only-if only.

So Mendelson had the shakes, Gibson reflected. Just like Ted. Some things never change, even after a decade of medical advances. He scrolled down nearly to the end of the file.

Only two of us left. I may actually get to like Donn …

Donnie asked me what he could do to help. I told him not to bother burying me. It was a good run. We had … nx … f … 09x>gaf2s

We've decid … try to make it back to the dome on foots. At lest they can fnd our bodees ther. & my diary.

Gibson was left with two mysteries: where were the two minirovers, and where were the bodies of Donnie Ramirez and Dave Culpepper?

It was all so tragic. How much of this did Hadley need to hear? She wanted to know, but she had her own expedition to run, and on its eve, he would let her sleep. The right time would come. He only had to wait for it.

V

Hadley got up early, before anyone else stirred. She opened her window shades to a blaze of white and blue. Out there, the day was clean, fresh. It had an expectancy about it. It was full of possibility.

This was the day she had anticipated for a decade, the beginning of the expedition, *her* expedition. Not everyone was coming. They had lost Joel and Orri, then Ted, whose body lay cold and quiet in the medlab. The thought saddened her, but she also felt anger, and then guilt for those feelings.

She begrudged the fact that Ted's death—and the others—could rob this important day of its joy.

Hadley Nobile decided to not let that happen. She would not let darkness steal away the triumph of her grand exploration. She decided to be intentional about it. This was a day of joy. This was her expedition!

Hers? For a moment, the thought seemed odd, inappropriate. Who was she kidding? Countless people had helped her, had donated time and money and resources. Still, it felt like her show. She had done the real work, had the vision, planned logistics and supplies and routes. It was her voyage of discovery. And if she could discover the other thing, the thing on her personal agenda, the thing that had nothing to do with Europa, that would be even better. If she was right, it would depend on the others.

But the expedition meant something else. As the team ventured into the remote wilderness, their living quarters would shrink from an international science settlement to a cramped outpost. She could feel the claustrophobia already. The noose was beginning to tighten. The possibilities of escape, or of simple safety, grew more and more limited as her expedition reached more remote locales. Now, in light of the menace among them, the idea of exciting work at a remote science outpost made her skin crawl.

—*—

To Gibson, breakfast seemed an odd affair. Six people sat at table, polite to a painful degree. Everyone seemed self-assured, poised, even confident.

More coffee?

The trip ahead would be challenging and dangerous, but these were professionals to a person. Each one had expertise, and each had training for the expedition. Still, the room was filled with a tension beyond the normal pre-trip butterflies.

Pass the eggs, if you would?

On the outside, calm. On the inside, turmoil. Who would be next? When? How much time did they have? God, how much *life*?

He toyed with the idea of putting it all out there, before the group. *Everybody knows what is happening here. Somebody is systematically murdering our colleagues and I want to know why.* But Hadley had put the kibosh on the idea. It would only heighten an already tense situation, she had said. So he waited. He didn't have to wait long.

"Well, that's that," Dakota said, tossing her napkin on the plate in front of her. "Great breakfast, but food is not what's on everybody's mind, and we might as well talk about it. Three of our friends will not be joining us at our dome in the south because they're dead. They're not dead of a virus or a heart attack or cancer, and we all know it. Someone is killing innocent people, and that someone is probably right here among us."

A pall fell over the room. The forced civility suddenly dissipated, giving way to controlled terror. It was as if someone had turned a light on in a room best left darkened. Amanda looked at Aaron. Gibson looked down at his plate; the deed was done, the cat out of the bag.

"How refreshing," Sterling said, his expression inscrutable.

Dakota stiffened. "Refreshing?"

He swirled the liquid in his glass, studying it. "Yes. Your view of innocence. There's a certain childlike virtue to it, an idea that justice should prevail, or that there should be something fair about life. We find it abhorrent that anyone should suffer 'unfairly,' or that the bad guy would get away with the crime." He put his glass down and looked up at her with a smile devoid of any warmth. "If you've done wrong, you'll be punished. If you're not guilty, you'll be left alone." He rested his elbows on the table and steepled his fingers. "The concept of guilt versus innocence is, of course, a philosophical construct. No one at this table is completely innocent, in the strictest sense of the term. We've all done things I'm sure we're not proud of, and a time of war brings those types of things to a head. Life has a way of forcing people into situations that are anything but innocent. It becomes a matter of nuance, of degree. Something a child would not understand." He said the last looking deep into Dakota's eyes.

"Curious word, innocent," Dakota shot back. "People can spin it in all sorts of ways to suit their needs." She can dish it out as well as she can take it, Gibson marveled.

"What would you suggest we do, my dear?" Amanda asked. "Have a kangaroo court? String up all our suspects? I'm sure everyone here has one or two skeletons in the closet."

"The problem is that someone's skeletons are tumbling out onto the rest of us." Dakota stood and turned to leave. In a tone devoid of emotion, she said, "See you all at the loading dock."

—*—

Benjamin Reynolds, the new med tech at Taliesin, pressed his nose against the glass of the observation deck. Pilot Dirk O'Meara stood at his side, taking in the departure of the largest south pole expedition Europa had seen since before The War. The rumble of the giant rovers rattled through the floor and into the walls, even through the vacuum outside.

"Those things are massive," The med tech said. "All this equipment. It's crazy."

"If you ask me, they're all crazy," Dirk said, nodding toward the glass. "I should know: I flew out with them for the last few months."

"I wouldn't go down there on a bet." The voice from behind belonged to the man with the hypermop. "That *is* crazy." In the window glass, Ben saw the man's halo of wispy gray hair reflected from behind. The hunched figure cleaned a path as he worked his way out of the observation area.

The expedition's two gliders slid by, skimming just half a meter above the ground. Each carried several months' worth of food, oxygen, water, and an ATV. At the back of the far glider, the ungainly submersible rocked gently beneath its tripod.

The convoy snaked south, carrying sophisticated science equipment, the latest in transport vehicles, the most advanced high-technology gear money could buy, and three new members. Jeff Cann, Tefia Santana and Felicia Tanaka rounded out Hadley's depleted crew with fresh expertise and new blood. The gliders left in the tradition of the ancient cavalcades, like a great desert caravan embarking upon the Silk Road. But the rubble paths of ancient Asia were a tame counterpart to the arctic wasteland that stretched ahead of this procession. Dirk wondered if he would see any of them again.

8

Dome in the South

I

This was it. Hadley could feel the mounting momentum with every meter they crossed, the inertia, the life-and-death commitment. And with each turn and shimmy and dip, she knew—more than ever—that there would be no turning back.

The ride was remarkably smooth. The gliders kept a clearance of 1 to 2 meters, skimming the surface. In places, the ice ponded in oval basins. But in others, the ground heaved up in jagged edges or piles of shattered talus. The morning Sun glittered across the vista. Crimson, purple, orange and blue flashed here and there, fireflies in the vacuum.

As Taliesin disappeared to the rear, the vehicles came upon a broad gray band. Here, the crust had fractured and drifted apart, leaving a frozen highway, its interior flooded as the ice walls moved apart, filled either with liquid water or slushy ice that froze in place. The band was ancient, as it had smoothed over. Fresher, more rugged bands lay ahead.

Sterling sat at the helm of Glider One, with Hadley riding shotgun. Dakota Barnes and the Grants rode in the rear compartment, an area twice the size of a minibus. No passenger had opted to discuss the subject brought up at breakfast. The matter had taken on an unspoken taboo. Hadley stood at the ready, in case Sterling and Dakota needed to be separated. *She would pull this glider over, kids.*

Gibson took the wheel of Glider Two, carrying the Taliesin contingent: Jeff Cann, Tefia Santana and Felicia Tanaka. "I want to get to know them better

© Springer International Publishing Switzerland 2017
M. Carroll, *Europa's Lost Expedition*, Science and Fiction,
DOI 10.1007/978-3-319-43159-8_8

before our field work," he had told Hadley, "and I want you to keep an eye on our survivors."

She hated his use of the term, but she had to face the fact that he was right.

"We should be line-of-sight with one of the sats soon," Sterling broke in on her thoughts. "Higher data rates. Too bad they lost the forth one of the constellation. A tech at Taliesin told me they used to have close to 24/7 coverage. As it is, I guess we wait."

"I can wait a bit," Hadley said. "But as soon as one of those satellites comes up, I'll need to do some research."

He shook his head with a smile. "Always so focused! Don't forget to enjoy the ride. We're seeing stuff that few people ever have."

She smiled faintly. Keeping an eye on the data rates, she glanced out the window. Toward the western horizon, a 300-meter high rise of burnt-orange ice marked one of Europa's famous domes. Sulfuric acid, magnesium sulfate and other assorted hydrated salts migrated through the disrupted ice from deep within the ocean below. There, radiation reacted with the chemical witch's brew, painting the ice in a palette Rembrandt would have loved.

Beyond the ice dome, ancient terrain undulated in a washboard texture. The corrugated surface rested atop scalloped cliffs 100 meters high. At the base of the cliff, slopes of debris lay against the ice walls, sloughed off from the cliff faces.

Hadley saw the upheaved ice as a symptom of great forces within Europa's depths. Tidal energy from the push and pull of Jupiter's gravity focused energy into rising plumes of warm ice. Ganymede and Io contributed by forcing Europa into a resonant orbit where Jupiter could have its way with the little moon. When the plumes reached the base of the cold, brittle crust above, they spread out like a warm subsurface pancake. As the warm ices sculpted Europa's surface, they left domes, depressions, and the famous chaos regions. She wondered at the variety of forms Mother Nature could present using only one form of energy. And what was exciting to her team members of the biological sciences was that even in places where the crust was thickest, material from the ocean made its way to the surface. What biological mysteries lay out there, just beneath the surface? Perhaps they would soon know.

As the cliffs retreated to the rear, the landscape became uneven, with rocky hills and hummocks glimmering in the sunlight. Ahead, a twin-peaked mountain of ice rose from the rough matrix some 250 meters into the ebony sky.

Sterling tapped the front console. "Picking up the first navigation beacon. We're right on the money."

"Nice that they plopped those relays down for us," Hadley mused. "Looks like you could get lost out here."

"No road signs; that's for sure. A bit like the interior of Louisiana. Ah, I see it out there. Just to the left of the big iceberg. It's a tall orange can on a tripod; see it?"

"Got it. How many more?"

"This is numero uno of six. They said they set up the last one in sight of the dome."

Hadley glanced back down in time to see a jump in the data rate. Now was her chance. "Our satellite must have risen. I need to go back."

"I'll just watch for reindeer and sleighs from up here," Sterling quipped.

Hadley left the front cabin and made her way toward the back of the rear compartment. Dakota looked away from the window, gave her a quick wave, and then returned to observing the view. She never missed much.

Aaron appeared to be sleeping. Amanda, across the aisle on the other side, seemed to not notice her. But as she made her way past, Amanda said, "Beautiful, isn't it?" She kept her eyes on the view. "Desolate, yes. Bleak. But unspoiled by poverty or crime or war. A chaste landscape, a tabula rasa waiting to be written upon by us."

"Let's hope we write something worthy," Hadley said, patting her on the shoulder as she continued on. She settled on a foldout chair against the back bulkhead and keyed her pad. She needed the seclusion to do her work. The connection was adequate, so she began to troll through the information. The International Science Foundation had given her special software for background checks while she was vetting her team, and she had kept a backup on the pad. Not something they would have loved, but not *strictly* against the rules. She had double-checked Sterling's vitae before departure this morning. Sure enough, his dossier declared his location to be on Earth up until just weeks before the mission. Might he have misspoken to Dakota? Perhaps she had heard him wrong or misinterpreted something. But Dakota was a detail person. Her scientific discipline made her a good observer, too. The most likely scenario was that Sterling had let slip something he didn't want anyone to know.

She scanned travel files and passports. Finally, the software zeroed in on a data point:

STERLING EWING-RHYS: DEPARTURE MARS 07 MAY 2157 11:45pm local

ARRIVAL INTERPLANETARY DEPOT PHOBOS 08 MAY 2157 3:18am local

Clearly, he had not come directly from Earth, as he had claimed. It was far more expensive for Earth flights to detour to Mars en route to Phobos than to fly direct. If he had gone to Mars first, he must have had good reason.

It was far more likely that he had been there for an extended period, rather than a stop-off.

She frowned and looked up from the screen. Out the window, the terrain smoothed and sank toward a dark low plain. To the west, a great stairstep of terraces bounded one side, while gentle slopes framed the darkened lake on the east. Sterling picked up speed as the surface smoothed even more, presenting a molasses-colored plain of undulating ice. Toward the center of the plain, the dark material had sloshed over the lighter terrain, painting tawny stripes up the flanks of several raised ridges. Clearly, the dark material had once been a flood of icy slush.

Only a scattering of craters lay across the field. Dark material had flooded a few, while others spread out in long lines. Hadley guessed that these were secondary craters from a much larger impact in the distance.

Sterling. She distanced herself, trying to be dispassionate about the reports. She liked the man. But facts were facts. As she pondered the contradictory records, another data point appeared from a second search.

VISA ISSUE: EARTH/MARS 2153
VALID THROUGH 2161
POINT OF DEPARTURE: EARTH
POINT OF DESTINATION: MARS

That was it. Sterling had left Earth nearly four years before the expedition set off from Phobos for Europa. He had been on Mars after The War, and for a very long time. He had lied to her.

She needed just one more part of the story, but she would have to dig deeper, and there wasn't time now.

II

Gibson rested comfortably in the driver's seat of Primary Glider Two's cab. In the seat to his left, Tefia Santana dozed. He supposed she had seen much of this landscape before. She had been stationed at Taliesin for three years or so, and had gone on several "outings," as she referred to them.

Gibson thought about the other glider and its precious cargo. In many ways, Hadley was his best friend. She was certainly his oldest. As she had suggested, he did feel protective of her. And there she sat in an enclosed vehicle a dozen meters ahead, riding with an unknown murderer.

His thoughts turned to Dakota. She stirred feelings in him that he had not felt since the death of his wife. She was exciting and spunky. He wondered about her life during The War, when she learned the disciplines of martial arts.

And he wondered why someone in college didn't hook up with her. He was assuming she hadn't, of course.

But she was so young. More to the point, she was a coworker. Science would provide more than enough distractions on this trip without some emotional romance getting in the way. It was probably a good thing that she was in the other glider. Some distance would be good.

"I remember those," Tefia said.

He glanced over at her. "Sleeping Beauty awakes."

"More like Grumpy. I hate naps." She gestured toward the window with her chin. "See those?"

Out the window, semicircular mounds rose from subtle depressions.

"Ah, yes," Gibson said. "The domes. Lenticulae."

"Lenticulae?"

He shrugged. "Freckles. Spots on the Europan ice. These brownish areas come in little gaggles, but they're a variety of things, like chaos areas, pits, depressions..." he pointed out the window, "and your domes. Early on, they just called them freckles. Lenticulae."

"Creative."

"You do get creative when all you have to go on is some fuzzy flyby images. But they figured it out later."

"They look like they're sinking into the ground."

"Actually, they are," Gibson said. "The domes do, anyway. Stuff comes up from underneath, punches through the surface as the heat from below softens the crust. But the ice can't hold up all that weight, so the domes of fresh ice begin to sink back down. That leaves those little moats around them."

Tefia nodded. "I see. They must be pretty strong. The dome over there has pushed up that whole ridge."

Gibson scanned the view in the direction of Tefia's gaze. Twin parallel ridges wandered along the flat plain. To one side, the ridges bent eastward, apparently shoved aside from the upwelling ice. Ghostly stripes crossed underneath the ridges, marking the scars of more ancient ridges, now sunken to phantom pathways of gray and tan. Rows of rolling hills and low domes stood out against the horizon.

Ahead, Gibson noticed a notch in the ridge. As the two gliders approached, he could see a wave of brownish ice that had broken through, burying a portion of the long cliffs and swirling around the base of others.

"Nice lobate flows there," he said absently.

"Now you sound like a geologist," Tefia said. "I was beginning to wonder."

"They remind me of some of the formations I've studied on Enceladus. In fact, these parallel ridges—if you scale them down—could be twins for the ridges in the south of Saturn's little snow-white moon." He stopped himself. Sometimes he found himself in lecture mode, even though those around him were simply showing courteous interest. Hadley had been the one to point it out, and she kept pointing it out until he broke himself of the habit. He turned to Tefia, trying a more conversational approach. "So, you've been this far south?"

"Only just. We had a rover wreck down here somewhere. I had to fly out to triage. It was a bad accident, but we didn't lose anybody."

"Nice work," Gibson said.

"Sometimes it's more a matter of luck."

"*Lucky* they had you there."

She didn't respond but gazed at the passing mounds.

Gibson, whose life's work had been the study of ice moons, focused on something else, an uneasiness that had been distracting him. What had he decided with Hadley early on, when she was going to meet with Ted? *Stick together. Don't split up.* And yet, they had split up. If someone wanted to bring harm to the rest of the team, now was the perfect chance. Everyone was in one isolated spot: Primary Glider One. It would have been far safer to split up the crew, to mix them with those from Taliesin who were, presumably, not targets. As it stood, they may have made things much easier for the murderer. In fact, Gibson realized, he was the only safe one now, the only one physically removed from danger.

He felt a wave of heat wash over him. His palms slid on the joystick over a gloss of sweat.

"You know what I always wondered?" Tefia said. "I always wondered why this place wasn't a big ice rink. Seems like the water has seeped out of the ground here and there. Why wouldn't we see nice flat ponds, smooth places like those mirrored lakes I used to skate on as a kid?"

"Where was this?" he asked, trying to hold his voice steady.

"Wisconsin."

"Well, for one thing, Wisconsin has air."

"Yeah, frigid air. Frosts your nostrils shut."

"When the water surfaces on Europa, it bubbles and froths in the vacuum. That's why you get all these spongy textures you see at the base of cliffs or out on the plains."

"Ah, I see." She leaned against her window. Her eyes drifted shut.

Gibson keyed his comm. "Primary Two to Primary One. Copy?"

"Four by four, Gibs," came Sterling's soft drawl. "How's life in your neck of the woods?"

"Things are smooth sailing." Gibson tried for a casual tone. "Is Hadley handy?"

"Very. Have you seen her repair a comm headset? But if you mean *is she here*, she is not. She's in the back. Anything I can help you with?"

In the back, with all of the suspects. "I just need to talk to her about something."

He glanced at Tefia, who was stretching like a lazy cocker spaniel just up from a nap. She stood and pointed in the direction of the bathroom, closing the cabin door behind her. Gibson was alone. Perfect.

"Probably not a good time," Sterling said. "She went back to do some research. She wasn't specific, but I got the feeling she needed some privacy."

Alarm constricted his throat. "Do you think you could get her on the line for me?"

"Is it urgent?"

Who was this guy, her bodyguard? Gibson's suspicions rose. "I think she might need to know what I have to tell her."

"Shall I relay the message to the back over the intercom?"

"Better to get her on line directly. It's personal."

Gibson's request was met with silence. It dragged on for a dozen seconds. "Sterling, do you read?"

More silence.

"Sterling?"

"Hey Gibs." The voice was Hadley's. Gibson blew out a long breath.

"Good to hear your voice, girl."

"Yeah, we've been apart for all of three hours. What gives?"

"Are you on a secure line?"

"Give me a second." The static of a moving microphone came over the headset. "Okay, I'm good."

"I hope you stay that way," he said. "I've just realized that you and I would be far safer if we had split up our merry band a bit more evenly."

"Yeah, I thought of that, too. But you know it's too late. We're on the road and we can't stop to shuffle. It would look suspicious, and we need to get to our radiation shielding down there."

"Suspicious is preferable to dead."

"There's safety in numbers," she said in a soothing voice. "Nobody would make a move in a close crowd like this. It doesn't fit the pattern. You just worry about keeping it between the lines and I'll stay surrounded by lots of people. We're fine, for now."

"For now," he repeated. But he knew he wouldn't relax until they made it to the end of their journey after a very, very long day.

III

Gibson rubbed his eyes. Tefia had taken over the driving for a couple of hours, giving him some needed down time. Now, he was back behind the wheel, positioned for the final push. The globe of Jupiter steadily sank toward the horizon as they pressed south.

"Passing relay number five," Tefia called out. "It won't be long now."

The glistening panorama around them morphed from shimmering plain to rugged terrain and back again. Hadley's voice came through the radio static.

"There it is … our little outpost!"

"Dome away from home," Gibson said as he spotted the shining hemisphere looming over a rise.

"We made good time," Tefia said.

"Do we have a parking lot?"

"Bear left. The airlock's on that side."

Gibson relayed Tefia's tip to Primary One. The two massive vehicles came to rest in a flattened hollow to the east of the main dome assembly. On the nearest side, a small airlock protruded from the base of the inflated hemisphere. Opposite, behind the dome, two full habitats extended directly away from it.

Gibson looked out the port side window at the other glider. His teammates filed through the hatch. He tallied them; all present and accounted for. He let out a sigh of relief, and noticed Tefia looking at him.

She cocked her head to one side, smirking. "I didn't take you for the mother hen type. All your chicks make it home?"

"Looks like."

_*_

The pressure dome reared into the sky like the arched back of a great whale, bulging with airy stresses from within. The distant Sun flashed against its billowed sides. Behind the veil, Hadley could just make out contours of equipment stands, colored lights, supports and cables. Jeff Cann was out of Primary Two now, headed for the airlock, carrying only one piece of equipment: a massive coffee mug, boiled empty in Europa's airless environment. She watched the others as Cann cycled the chamber. Sterling stood near the hatch. She kept an eye on him.

Hadley switched to the common channel as the group made their way toward the open door. Tefia was speaking.

"… have to go in in two groups. The lock's not big enough for all of us. Come on in."

The Grants, Gibson and Dakota were closest, so Tefia took them in with her. The hatch closed, the air cycled, its inhabitants went on into the outpost interior and the lock cycled again. Hadley moved closer to Sterling as the remaining group entered the cramped airlock.

"Nice and shiny," Felicia Tanaka said. "Looks brand new."

"Just give it a couple days," Hadley quipped.

"Yeah, this is our mudroom," Sterling said. "These places don't stay so clean."

The pressure pumped up and everyone unsealed their helmets. The air smelled like a new car. Polished plastics, chrome and blue plexi gave the appearance of a freshly unwrapped toy on Christmas morning. The far hatch opened and Gibson gestured to the group, bobbing like a butler.

Hadley let the group file on past Gibson. She paused, leaning over to him, and whispered, "Sterling lied to us."

His eyebrows bounced up his forehead, but he said nothing as they walked on. The dome towered above them. Ground level opened into a sweeping floor space. Despite the crates and racks of equipment, the area felt spacious. Hadley had studied the initial floor plans, the logistics diagrams, the electrical and plumbing charts. She had scrutinized most of the nuts and bolts of the structures and signed off on the design of the outpost. But in person, it really was impressive. Tefia called from the far side.

"There's no room-to-room interface here, so everyone will have to keep in touch with your wrist comms."

"Yeah, or just holler down the hall," Dakota said. "The place isn't that big."

Tefia continued. "The outside access bay is over here, for offloading large cargo and our sub."

"Excellent," Hadley said. "We'll leave the ATVs outside. Let's unload so we can relax."

Jeff Cann bustled through the small crowd. "I'm off to bring the reactor online. As soon as we're unpacked I'll fire up the radiation shield."

The overhead dome looked like the inside of a hot air balloon. Its clear fabric, misted by the interior humidity, billowed against a Fulleresque web of cables, undulating in the stirring air like a grand, breathing thing. On the far wall, another lock led to the living and laboratory areas in twin large habitats. Beyond them spread the processing plant with its heat and air exchangers. That must have been where Jeff Cann was headed.

After a moment, Cann ducked back through an access hatch to the side of the entrance. He brushed his hands across each other. "All set."

Amanda stood with arms folded against her chest, gazing up at a structure twice her height. "Those things look like oil drums on steroids."

"Or containers for gigantic doughnuts," Dakota added. "Please, *please* let them be giant doughnut containers."

"No such luck," Cann said. "That's our spool."

Dakota frowned. "Spool?"

"It's full of xenofibre."

"That huge thing is full of xenofibre?" Awe suffused Dakota's voice.

Cann nodded. "'Bout 27 kilometers' worth."

Amanda said, "Must be worth the mines of Solomon."

"Only the best for us. First we send the ROVs, and then you all get to go." Cann stared at a circular metallic hatch lying against the floor. The impressive port spanned 4 meters, just large enough for the torpedo-like submersible.

"Under there?" Sterling said, his eyes glued to the access.

"Yep, that's our doorway to the unknown, brother."

IV

The habitats consisted of two chambers and a small antechamber each. The advance team had outfitted the first chamber as a laboratory to study *in situ* samples, while the second was laid out as personal quarters. Hadley mentally checked off the names on her list as they settled into their living spaces. The Grants took the room next to the laboratory. Aaron immediately set to work arranging a small medlab for emergencies. Hadley, Dakota, and Felicia took the first room in the second habitat, and the second room accommodated Tefia, Gibson and Jeff Cann. Intrepid explorers all. One of them a murderer. She found herself missing Ted.

Hadley settled into her bunk. One of the Europa sats was still in range, but just. She pulled up her files on The War, the ones that seemed to intersect with Sterling. She was getting pretty good at sifting through the fragmented war records. Before this trip, she mused, she would not have had the patience. Most people didn't. Now, she had motivation.

She found a source that was a goldmine of records—electronic accounts from the web just before it all collapsed, and a shorter section of records from just after most of the infrastructure had been recovered worldwide. She returned to an account she had saved from her earlier trolling.

Large groups of former Eastern Alliance members are under suspicion of war crimes, but with relaxed laws governing the issue, and lagging interest for supposed justice, many will simply go free without what many consider to be due diligence. A partial list of the Antarctic and Asian arenas was released by the War Crimes Tribunal in hopes that [DATA MISSING].

Hadley scrolled down past several headings: Asia: CAMPS; Asia: LOGISTICS; Asia: SUPPORT; Antarctica: LOGISTICS [DATA PARTIAL]; Antarctica: CAMPS.

That was what she had been looking for. The list was partially corrupted, but she sifted through what was there:

Chelsea, Darlienna . . . . Camp Commandant in charge of prisoners
   Delvin, Walter . . . . Camp Commander
   DeVeres, James . . . . Medical Pl—
   Diw- - - . . . . ical
   Enteswide, Roguet . . . . b—
   Evashtar, Erin . . . . Medical Chief, alleged member "Seven Sisters"
   Evelyn, J - - . . . . ief, support
   Ewing, Sterling . . . . .transport

There he was. Either his name was partially missing, or he had been going by another at the time...or he was going by another now for protection. What did they mean by *transport*? She flipped to the terms section.

TRANSPORT: Involved in transportation of goods or people to the camps themselves, often pertaining t - -

Damn that lost data! But the account had given her enough. And what it provided chilled her to the core.

—*—

The morning saw frenzied activity as Jeff Cann brought the submersible in. The sub entered the domed warehouse amidst great fanfare, an unlikely hero to conquer Europa's ice shell, a new Nemo to plumb the depths. The candy-apple red torpedo, finally released from its stand, swung gently free beneath the crane.

The submersible rotated slowly, a humble curtsy before her adoring audience. Her rear propellers gleamed, her glass observation ports glistened, her tools stood at the ready inside the indented belly.

The group ferried other supplies through the airlock and loading bay throughout the day. The vast floor space under the dome didn't seem so open any more. Hadley oversaw every aspect of the offloading and stowing of equipment. The sight of her crew working together gratified her. *Why can't we run life like this? Not just here, but in the worlds across the Solar System?* It seemed a naïve pipe dream. Perhaps Amanda had been right: Europa was different, an Eden-like place untainted by what Ted had called "fallen" humanity. Was there a new Garden of Eden beneath the white vista out there, perhaps huddling around a volcanic vent on the ocean floor a hundred klicks below? If so, she would be the one to find it. She would be the new Eve, and it would be up to her to decide what to do after that.

V

"Just a little safety tip," Tefia said to the assembled group. The central dome was the only place large enough to convene anything like a group meeting. "If you go outside, stay inward of the blue markers. Beyond that, our radiation shield tends to hoard the incoming radiation. It's great for us in here, but not so good if you go out too far. When anyone needs to leave site, Jeff will need to shut down power to our mini-magnetosphere so that energy will dissipate. Good?"

Everyone nodded.

"Okay, group," Hadley said. "Good tips from the pros. Now, Gibson will be briefing us on our upcoming activities. Gibs?"

"I've got our tentative schedule all worked out. It's not written in stone—"

"I thought that's how all geologists wrote," Dakota jibed. Everyone laughed.

Gibson grinned and continued. "After we do our recon with the biobots, our teams will go down in pairs, with one person in the role of researcher and one a technical pilot familiar with the hardware." Jeff Cann nodded. He would be doing the heavy lifting in the piloting arena. Gibson continued. "If you'll all pull up file AG-2 on your wrist-tops and follow along…" Everyone looked down at their personal monitors. "According to the records from Taliesin's set-up team, the shaft goes down at a 30° angle for just under 3 kilometers. Then it hits a lens-shaped false sea, what we call a Schmidt feature. That's where we cut the sub loose from the cable. It should be smooth sailing for another klick or so until we reach the bottom of that underground lake, and then we pick up the shaft again."

"How do we get the sub down through there?" Dakota asked.

"It's all filled with water from that first Schmidt lake on down. Once we're submerged, they cut us loose of the cable and we can go where we want. The cable stays in the shaft, acting as a relay antenna. We descend another 12 kilometers and then hit two more Schmidt features, one on top of the other, that have collapsed into each other."

"The term collapse does not fill me with a sense of security," Sterling said.

"They turn out to be pretty common down there. But we will need to keep a close eye on the walls. Probably some rough ice on the sides. So, the two Schmidt lakes form a really deep fused waterway, but we aren't quite there yet.

We need to descend a dozen kilometers through the open water of the twin Schmidt lakes before we come to the last bit of solid ice, only about a kilometer thick. We pick up our shaft there again. Then, it's open sea below."

Hadley stood up and said, "So here's the strategy: The first team will reconnoiter the passage down, get free of the ice and do a survey of the upper sea just beneath the ice. Amanda, I'd like you on that one as first responder to any biological detection. Second team departs the next morning after we've had a chance to debrief and do contingency sampling. This will be a benthic dive, way down. I'll be on this voyage to the bottom of the sea, to look for active volcanism, biota around seafloor sites, and so forth. No matter what I find, Dakota will be on the next dive. Good?"

"The biology team is ready to go," Dakota enthused.

"Yes," Hadley said firmly, "but it's all contingent on any geyser activity. That will trump our schedule."

Jeff Cann put in, "In the meantime, I've started remote recon with the biomimetic subs. The first two are through the upper borehole, and one is ready for final entry into the ocean."

"And I've gone bionic, thanks to Tefia's implants," Sterling tapped his skull.

"Really?" Felicia grimaced. "Let me see."

Sterling leaned over and parted his hair at the side. "See?"

"Does it hurt?"

"Only when I do cartwheels. But look, it's a perfect fit."

Sterling placed a headset over his head, carefully docking it to the two ports in his skull. "I can now be one with my robots."

Felicia shivered.

Cann crowed, "It's time to get this show on the road!"

"And nearly time for us to hit the highway," Gibson said. "We expect activity at any moment now. Earthquakes are on the rise." He exchanged a glance with Hadley.

She turned her head to scan the group. "Okay, team, we'll take a break, get our last-minute business done tonight—"

"Like writing our wills?" Sterling said.

Hadley smirked. "Should of thought of that before we left. And as I was about to say before I was so *rudely* interrupted, the geyser group will prepare for departure on a moment's notice, as soon as Gibson's instruments tell us it's time. If things are still quiet in the morning, we dive. Let's maneuver."

—*—

Sitting next to Amanda, sharing a small monitor, Hadley felt cozier than she had in months. When she relaxed, when she was away from Aaron's side, Amanda could be quite companionable. Hadley thought again of her grandmother, and of the talks they had shared on the porch swing at Gramma's house. Amanda tapped her screen, speaking in quiet tones.

"It will be fairly straightforward to carry out a microbe survey using Remote Operating Vehicle Two. Or is it ROV3? I get them confused."

A voice came through the open door. "ROV3 is the one that looks like the glassfish—knifefish; whatever." Hadley and Amanda glanced up. Aaron stood in the doorway, waiflike. He wasn't that small of a person, Hadley reflected. Just average. But for his size, he held an insubstantial quality, as if something had hollowed him out and left him a mere shadow.

"Yes, that one," Amanda said. Aaron wandered into the room and came to rest on the other side of the table.

"That's good," Hadley said, hiding her disappointment at Aaron's intrusion. No time for personal talk now. "You can get started as soon as the ROV is ready."

Amanda pressed her fingertips together. "I've got a battery of experiments all ready to go."

"What's on tap?"

"New camera, for one thing. I'm very excited about it. It's essentially a portable electron microscope. I can see structures the size of organelles in a cell."

"That should come in handy," Hadley encouraged.

"Let's hope. Then I've got a PCR sequencer to search for DNA. Of course, any Europan life may not be based on any DNA we can recognize, but there's a theory that the DNA structure, or something similar to it, is universal to life, so there we go."

Amanda paused her soliloquy and smiled at Hadley. "PCR? Polymerase chain reaction sequencer?"

Hadley frowned and shook her head.

"You don't have a clue what that is, do you?"

"You're talking to a geologist. Biology is above my pay grade."

"Fair enough. A PCR array is a thermal cycler. It multiplies a single segment, or a few segments, of DNA. Generates millions of copies so we can see them and manipulate them for study. For example, we can search for a TATA box, something that every DNA strand has at its beginning. Or we'll search for replicase."

Hadley nodded. "Ah, I see." She remembered saying something similar to an insurance adjustor after a robocar accident, nodding in the wake of the adjustor's barrage of information. She didn't understand much of that, either.

"However," Amanda pointed to the ceiling, "if life is not based on DNA but still makes use of amino acids, which in my book is more likely, we've got an amino acid analyzer. Here's the thing with amino acids: we find dozens in meteorites, so we know the universe is seething with them. On Earth, life makes use of about twenty out of the hundreds we find in nature. Ten of those used by terrestrial life also appear in meteorites, and our hunch is that these particular ones are used by life everywhere in the universe."

"A sort of lingua franca of the cosmos?" Hadley offered.

"If you will. Another ten in Earth life were apparently invented by evolution."

Ted would have said designed, Hadley thought, but left the idea unspoken.

"So it's highly unlikely that these would be the same in separate water/carbon life elsewhere. Remember that amino acids twist, and life on Earth uses the left-handed twisting version. But all right-twisting should work for life, too, and if we found all the amino acids were right-handed, that would be an indicator that the life was not based on the same source as that of Earth's—and maybe transported by meteor impact, for example—but rather came about independently."

"Europa's a long way to come on a meteorite," Aaron grumbled.

"That it is," Amanda said. "But we'll be looking for life-like lipids with another device, and I've got a powerful organic analyzer to detect the carbon-based compounds that we're not even expecting. It's got a mass range up to several thousand and a resolution down to about one millionth of an atomic mass unit."

Hadley swooshed her hand over her head.

"All you need to know is that it's good. If there's something in Europa's oceans, we'll find it."

"Even though no robots have found anything yet," Aaron added.

"Forever the killjoy, my dear. It is ironic, though," she agreed. "So many of these instruments have been here before, on precursor missions. We already know Europa's waters are rife with organics, seeded with the rich flotsam of prebiotic matter. But where is the life?"

"If there is any," Aaron said. "Maybe it's just a global chemistry experiment. Maybe Earth and Mars are it, and biology never got started here. And Martian life may have been seeded by meteorites from Earth. Maybe Earth is the progenitor world in this Solar System. Lots of lights on, but nobody else at home."

"Conversely," Amanda said firmly, "maybe life has taken hold wherever it can, wherever the building blocks are common: Mars, Europa, Enceladus, even Ganymede."

"Optimist. Besides," Aaron argued, "I thought you said Ganymede's ocean was trapped inside the ice crust, cut off from all that rocky stuff that your precious microbes would like to eat,"

She patted the back of his hand. "But not isolated. Currents of solid ice move from the rocky core up to the surface. Takes a while, but there's still delivery of minerals throughout the ice. Life may not be such a long shot there."

"Maybe you'll be the one to find out," Hadley said. "Once we have humans under way, Aaron and Tefia will monitor vital signs."

Aaron nodded. "Good." He said it simply, succinctly, but for the first time on the trip, Hadley saw something there that she hadn't seen before. Aaron was being protective of his wife.

Sterling Ewing-Rhys stuck his head through the hatch and looked at Hadley. "Hey, sister, Jeff says the submersible is all checked out and ready for tomorrow. I've got the human/robotics interface set, so Dakota can do some remote sampling while the initial crew is underway. She's already done some work with Jeff and the ROVs. Amanda, I envy you. Guess you've got the first passenger ticket."

"Indeed I do," she said with a little too much enthusiasm. "I'd better get a good night's sleep." She stood and left the little room, with Aaron silently following in her wake.

Sterling looked at Hadley. "Well, here we go. It's all in motion now. Excited?"

"Yes," she said, hesitating. "We're all going to be very busy from here on out."

"Cross your fingers that we don't spend all our time troubleshooting and repairing. Some of this high-tech equipment is a bit finicky."

"I think things are well in hand with all of our illustrious experts." She considered confronting him on the spot, but then thought better of it. She should wait for Gibson. He would be her backup. Suddenly, she realized she was alone in a room with a prime suspect. She tried to read him. His expression was inscrutable, not friendly, not hostile. He seemed to be studying her.

Sterling was about to say something when a rap on the door interrupted him. "Anybody home?" Felicia Tanaka entered and motioned to a chair. "Hey Hadley, Sterling. May I?"

"Please," Hadley said brightly.

"Just thinking about ice geology," she said.

"That's way beyond my focus," Sterling waved. "If you all will excuse me?" He nodded, almost a little bow, and ducked through the door.

"What's on your mind, Dr. Tanaka?"

"Felicia, please. I've been studying some of the orbital imagery in our proposed geyser region. Judging from features we've seen in the Tiger Stripes on Enceladus, I'm putting together a route map. If you look here ... and here and there ... " she tapped on her tablet, "These outflows are quite porous."

"Like a hard sponge."

"Flash-frozen bubbling water, yes. Typically, we get these areas on Enceladus adjacent to active vents. I think these three sites are our best bets. If we don't see direct activity, we can at least set up on this promontory as our base observatory."

"Yes, good. And once things get going—"

"We deploy our remote stations. The ones we've got can clock the velocity of particles, their chemical signature, size, the whole nine yards."

"And you've got those loaded?" Hadley asked.

"Not yet. I'll have them packed by morning."

"Can you use some help?"

"Love some."

"I'm sure you can find someone." Hadley paused. A grin broke across her face. "Just kidding. Let's go."

VI

The submersible seemed so much larger when it was hanging in the bay at Taliesin. Here, cocooned in a sling over the borehole, the red torpedo looked small and timid, especially for two people. Unlike Ted, Jeff Cann had never taken this particular craft out, but it bore close similarity to ones he had piloted, and he'd spent plenty of time in the simulator. He wasn't concerned for himself; he was worried about his passengers.

Cann leaned close to Amanda, helping her to adjust her dive suit. Even though they would be inside a vehicle, their garments carried webs of heating radiators, medical monitors and other technological wonders to make the voyage safe and fairly comfortable. Now was the time for care. Operating at depth could bring out the worst in people. It had a magnifying effect, a multiplication of phobias and weaknesses. He had seen it before. He didn't ever want to see it again.

"Are you claustrophobic?" he asked her.

She glanced at the sub, and then back to him. "Not so much."

"A little?" He cinched her collar, hard.

"Sometimes."

Jeff looked across the room at Tefia and Hadley. "Then I'd like to request that you take a dose of mild sedative. It won't get in the way of your work. I've had the stuff. It doesn't even slow you down, in terms of cognition."

Amanda looked skeptical.

"It's a request at this point." The implication was that Cann could turn it into an order.

"It's your ship, Captain."

Jeff called out, "Tefia?"

Tefia held up a hypo. "Got your cocktail right here. You'll be much happier, Amanda."

Twenty minutes later, Jeff Cann called for them to seal the hatch. He lay on his stomach, surrounded by readouts and glass-domed windows. Amanda lay immediately behind him at a slight incline. Her hammock cradled her such that her head was above the back of his knees. At her side and above her head, a battery of screens provided data on all her experiments. Everything but the spectrometer remained flatlined until deployment in the water.

Cann powered up the sub's interior environment systems. "Here's a little breeze, Amanda. You okay back there?"

"The air is nice," she said.

Cann keyed his microphone. "Standing by. How do you guys read?"

"Loud and clear," came Hadley's voice.

"Amanda, try yours."

"Dakota, how do you read?"

Dakota's voice sounded garbled, as if she was blowing bubbles under water. Cann and Amanda could hear Hadley's voice in the background. "Stop that!"

"Sorry," Dakota said. "It's an old party trick. I can do a mean impression of the captain of the *Titanic*. Anyway, reading you four by four. Ready to monitor all biological experiments." She said the last like a game show host. Cann could almost hear the grin on her face. He would have to see about a deeper relationship. Later.

"Stand by for descent," Sterling said. The submarine shuddered, then slid smoothly down the sloped wall of the borehole.

The outside light blinked out as ice surrounded them. Cann could hear Amanda let out a breath.

"How you doing?"

"Good." A quaver infused her voice.

"We'll be sliding along the ice for a while, now. Think of yourself as an Olympic luge athlete." He gave his voice as much cheerfulness as he could. Would she be okay?

"Three kilometers seems like a whole lot of ice."

"It is, but it'll go fast. They're lowering us at about 5 kph, so we'll hit the water in less than an hour."

"Good," she mumbled. "Good."

Cann looked ahead into the darkness. He turned on the submersible's headlights briefly. The view ahead reminded him of the great Eisenhower tunnel in the Rocky Mountains. But this tunnel didn't have pavement or lane markers. Layers of color passed by, records of past epochs. The ice had laid itself down in cryoeruptive events, surface floods, melting and remelting, and tectonic folding from the inside. Rich blues and ruddy tans painted the neutral gray walls in undulating patterns. The marbled bands lent a visual cadence to the voyage, a repetition that mesmerized, hypnotized.

Amanda, too, seemed riveted to the scene outside. Now and then she would exclaim. "That red spot was intense! Wonder what caused *that*? That brown bit is the classic color of organics … That's the closest to a blue sky I've seen in ages."

The paying out of the cable slowed. Suddenly, the sub hit a floor of water. "Contact with liquid," Jeff Cann called into the headset.

"Roger that," Hadley said. "Cable release."

The craft rocked back and forth as it stabilized on the water's surface in the cramped confines of the bore shaft. Cann hit his touchscreen and the submersible began to descend, gently, farther into the water. "Clear of the shaft. Beginning descent." He turned to Amanda. "We're in the first Schmitt feature. They'll leave the cable attachment here and we'll reconnect on our ascent back. We'll sink for a kilometer and then pick up the shaft again. That section is vertical, so we can make a faster descent through the next section of ice."

Cann grabbed a joystick. The small vehicle responded rapidly to his movements. It pirouetted around and headed down into the abyss.

"Open water," Amanda marveled. "Time to start sampling."

"You go, girl."

VII

Sterling cast a wary eye at the gigantic spool as it cranked to a stop, suspending its kilometers of xenofibre cable below it. Without taking his eyes off the spool, he told Hadley, "Nothing much is going to happen for an hour or so. Then they'll enter the second set of Schmidt features. Shall I call you then?"

Hadley wasn't ready to leave, but Gibson jumped at the idea. "Sure," he barked. "I want to take another look at the seismographs, watch for action."

"As if there's not enough action down there," Hadley said, gazing into the bottomless pit.

Gibson shrugged. Hadley told Tefia, "Take good care of them. Call us if anything—"

Tefia held up her palm. "I've got this. Both of their levels look nominal. They're doing fine. Don't worry."

Hadley followed Gibson into the makeshift lab.

—*—

Gibson hunched over his monitor, studying a set of lines. There had been something, he was sure of it. A telltale jump in the seismometer data, a subtle hint of activity to the west. Could this be a precursor to the long-awaited geysers? Hadley leaned in urgently, keeping her voice low despite the fact that everyone seemed to be off in the loading bay. "This thing with Sterling just keeps getting better and better."

"What are you talking about?"

"Aside from his mystery Martian escapades, I made a disturbing discovery. Sterling may have been involved in more than Antarctic warfare. He was in the region of the camps of the Seven Sisters."

Gibson looked up from the screen. "Oh?"

"It looks to me as though he actually *ferried people* to those camps."

Gibson grimaced, again scrutinizing the complex squiggles. "Hitler had his Mengele. Al-Cunelan had the Seven Sisters."

Hadley nodded, thinking out loud. "*Planetary Geographic* said something about…how did they put it? 'Medical experiments with a dozen times the malice of the Third Reich's.' And somehow Sterling was in the middle of it."

Gibson tapped the screen once, then once again. Light flickered across his face. The lines resolved into more detailed rises and valleys. "It leaves a bad taste in my mouth, but much as I hate to say it, his war record shouldn't get in the way of his work here. We're all trying to bury those hatchets."

Gibson's comment reminded her of Ted, of an exchange she had had with the man of the cloth. They had been in the galley, just the two of them, waiting for the others to assemble for some meal—which one she couldn't remember—when Ted started talking about The War, about how people were faced with primal decisions in the worst of times. They got interrupted, but he had messaged her later to finish his thoughts. She was glad. Ted's words struck a chord within her, as they so often did. *You see, Hadley, a small cruelty, well-placed, a subtle envy or a demonstration of animosity that seems—on the surface—nothing more than petty, can poison a friendship. Or a community. Or an entire nation. Human lives are bound to each other with just that intensity, whether we see it now or much later, magnified. But the good news is that in like way, even the smallest good turn—just words of courage to someone starving of hope, the simple praise that provokes a smile—these things ripple across the great pond that is life. They influence the life not just of one person, but of generations to come. Each time it is passed along, it grows in generosity and kindness and selflessness.*

Those kinds of acts, the ones that heal and build and nurture, those acts give power to those who would make civilization rise again from the ashes of war. Each of us, every person, adds brush strokes to the painting. Will the end result be graffiti on a crumbling bridge, or a lyrical canvas of beauty?

His rhetorical question still hung in her mind.

"But clearly," Gibson broke in on her thoughts, "we need to find out about that missing part of Sterling's timeline."

"Yeah, we do. I wonder if we should take a more direct approach."

"As in?" His eyes didn't leave the monitor.

Hadley shrugged. "We could just ask him. After my dive."

9

Vital Signs

I

Jeff Cann maneuvered the little red submersible with care. He took pride in his navigation of the alien sea. Of any sea. He envisioned himself as piloting with the same flair as a Michelangelo sculpted a piece of marble or Beethoven composed music: nuanced movement, conservation of power, graceful turns and gentle velocity changes.

No light filtered down this far, but the headlights scattered beams of impossibly blue light. A smattering of particles glimmered in the artificial illumination, golden flecks drifting in expanding cones spreading before them.

Pings issued from the control panel.

"What's that?" Amanda snapped.

"We're picking up the transponder that the robots left at the entrance to the next tunnel. It's resting on the floor of this Schmidt chamber. Almost there."

Soon, they could just make out textures in the foggy light. The textures resolved into a ribbed plain of ice. And just in the center of the beams, a dark oval sprawled across the glistening floor of the Schmidt chamber. Next to the black entrance lay a small orange and black packet, the transponder.

The descent through the lower borehole proved hypnotic. The ship rocked gently as it passed by more of Europa's icy layer-cake walls. Finally, the tunnel opened into a vast, black emptiness.

Amanda took in a sharp breath. "Oh, here we are!"

"Davy Jones' locker got nothin' on Europa," he said cheerfully.

© Springer International Publishing Switzerland 2017
M. Carroll, *Europa's Lost Expedition*, Science and Fiction,
DOI 10.1007/978-3-319-43159-8_9

The vehicle steered a steady course as Amanda played a symphony of research on her instrument platform. Data streamed into the sub and on through the waters above, to the team waiting topside.

"Organics. We've got complex organics!" Amanda called. "Looks like some related amino acids. Nice."

Cann gazed into the darkness. Did he see the undulating forms of swimming things out there? Was his little ship sharing the alien abyss with great serpentine life forms, creatures never dreamed of, beasts that hearkened from an entirely new biological journey tracing back to a time when Europa was an ocean world next to a giant planet still glowing from the heat of creation? The idea charged him. The hair stood up on the back of his neck.

Amanda punched a few spots on her touchscreen. "Got some good microscopic samples. I can't wait to get this stuff back to the lab."

Cann's console lit up like a Christmas tree. The sub listed and hesitated. It began a slow rotation to the right. He called back to Amanda. "I'm going to backtrack a little. It's just about time for us to return." He hoped she wouldn't look at her chronometer.

He slowed the vehicle. It stubbornly continued its lazy spin. He shut one of the engines off.

"Are we stopping?" Amanda asked.

He remained silent, trying to figure out how to tell her in a way that wouldn't panic her. A wild engine was bad enough in 5 meters of water. But in an alien ocean dozens of kilometers down, it could well be deadly. "I'm going to be using our maneuvering fans to back up a little; give the main engines a rest."

She grunted and returned to her science. He was glad she was distracted. With the shutdown of the main engines, the small attitude propellers wouldn't last long. But somewhere above them, the borehole awaited. If he could get to it and line up, he could command the sub to float up through it, all the way to the cable. It was a long way. If there were strong currents anywhere in those Schmidt lakes, they wouldn't make it. He began to sweat. He turned up the environment fans.

"Can you go back the direction we just came from?" Amanda asked. "I want one more little run there before we leave."

He shook his head in frustration but controlled his tone. "Actually, our engines are a bit, ah, overheated and I don't want to tax them too much."

"Sad. But I assume we'll get another chance."

He sincerely hoped so.

Using the ship's buoyancy, he ascended without engines. The beacon was just above and to starboard. He coaxed the sub using the weak attitude props. They whined under the stress.

The pings came faster now. Amanda had become very still. He said, "Almost to the transponder now."

The underside of the ice layer loomed overhead, closer and closer. He looked for the entry point. Where was it? He flicked on a floodlight mounted atop the ship. Suddenly, mercifully, the ice surrounded them on all sides. They were ascending the lower borehole.

"Got it. Now all we have to do is rise through the two linked lakes and get to the upper tunnel, and they'll reel us in on the cable."

"You sound relieved. Are you relieved, for some reason?" Amanda's curiosity was clearly aroused by Cann's unorthodox ascent.

"Just having few issues with one of the engines. Nothing serious," he lied.

He radioed topside, putting the communication through his earpiece so Amanda couldn't hear. Sterling picked up. "And how is our good Captain Nemo?"

"We're ascending passively. I say again: we are ascending passively. Hoping to reach the cable without power. Will advise."

"What the hell's going on?" Sterling demanded.

"More on that later," he said with exaggerated cheeriness.

"Oh, got it."

Cann smiled. Sterling was a sharp guy.

The rise through the lakes stretched on interminably. It was Cann's turn to become claustrophobic. This sub wasn't big enough for him and a panicking biologist. He heard a soft sound that he didn't recognize at first. Was it coming from the side engine? Perhaps a short in the electrical system? Was it possible that the craft had drifted enough to be scraping against an ice wall? Finally, he realized what it was: Amanda was snoring.

The ice ceiling approached quickly. Cann slowed the ship as best he could and radioed up.

"Sterling, can you give me a few pings through the cable? I'm in search mode."

Amanda yawned. "Boy, not much room to stretch in here. I had a nice nap. Can you imagine?"

"You didn't miss a thing. Oh, except there was that Plesiosaur. Maybe you'll catch it next visit."

"I'll be sure to keep an eye out. Where are we?"

"Almost to the upper tunnel," he said, but he could tell they were considerably east of it. He turned the craft and began to back away in the direction of the cable's pings. His console gave a proximity warning. He tried to slow the ascent, but the ship seemed to have a mind of its own. It gave him a collision warning. Almost immediately, the sub bashed into the ice, bounced once, and floated back up to rest against it. The cabin lights flickered, then steadied.

The bumpers on top scraped along the ice. The sub was still moving, sliding along a slight incline in the frozen ceiling. Suddenly, it tilted sideways, rotated nearly upside down. Amanda let out a small "Oh!"

The sub listed the other direction and slid up into the tunnel. "We're there!" Cann called. He commanded the waiting cable to dock with the ship, which it dutifully did. "Sterling, can you get us out of here?"

"Ready to roll. Going up takes a bit longer, but we'll see you soon. Everything all right?"

Now that they were safely in tow, he could let everyone in on it. "No, everything is not all right. I had some kind of partial failure in the starboard main engine. I had to basically float up, and I've probably just about burned out our maneuvering jets."

"Roger that. We'll fix you right up when you get back."

Cann hoped so, for the good of the team. With this dive, they had just dipped their toes into the great Europan abyss.

II

Sterling was as good as his word. The next morning, he declared the submersible seaworthy, tossing out a few technical terms for what had afflicted the little vessel yesterday. Jeff Cann and Hadley donned their suits. All hands assembled to wish them a *bon voyage*. Cann crawled into the cherry red torpedo. Hadley waited at the hatch for him to give her the all clear to enter. She gave Sterling a thumbs-up. He nodded slightly and turned his attention to the great spool of xenofiber. Over his shoulder, he said, "This little ice ball must be a volcanologist's dream. Volcanoes above and volcanoes below."

"You got it, but no one has seen the ones below with their own eyes. Just ROVs, and all that footage was pretty fuzzy. Too far away, except for one that we sent down to a seafloor chimney. I'm afraid that one got cooked before we got it back."

"Don't you dare do that with ours," Sterling said with mock sternness. "I'm getting attached to my cherry hot-rod here."

"I'm a conservative researcher," she assured him.

"I guess I'm not real clear on the whole submarine volcano thing. I understand geysers. You get cracks in the ice, and the ocean wants to blob out into the vacuum. But undersea stuff? I'm just a robotics geek, but it seems like my high school geology taught us that Earth's sea volcanoes are all along plates or some such thing. I know Europa hasn't got tectonics, right?"

"Not like terrestrial ones. But for a reason, all you have to do is look out the window. Not here; can't see Jupiter from here. But at Taliesin you get a grand view of not only Jupiter but ... "

"Io," Sterling filled in. "Volcanoes. But it's a rock moon." He shoved his thumb toward the window and the ice outside.

"Don't think of the crust," she said, "but the core. Beneath all that hundred klicks of seawater, Europa's got a rocky core that's pretty big. And it suffers from the same cosmic tug o' war that triggers Io's volcanoes. Io's and Jupiter's gravity on one side, and Ganymede's on the other, push and pull enough to heat up the core. Nowhere near as much as on Io, of course, but all the same, that heat has to come out somewhere."

"And that somewhere is where you're headed."

"Let's hope."

"You won't be headed anywhere until you climb in," Jeff Cann said. "Come on aboard."

"Be gentle, Sterling. It's my first time." Hadley ducked through the small opening and unfolded herself on the hammock-like bed inside.

Gibson clanged the hatches shut, front and rear, and knocked on the sub's side. "All clear."

—*—

"No worries, boss," Jeff Cann said to his single passenger. "I can make this trip in my sleep now. We'll be through the ice tunnels and into open water in a jiffy. Well, jiffy may be an overstatement. Volcanoes, here we come."

"I can hardly wait," she said, gazing at the lines of ice rushing by the porthole. She couldn't help but smile at her pilot's enthusiasm. "You seem awfully happy."

"Doing what I love best: sailing the mystery seas in my little submersible. It's a good life."

"Yes, it is," she mused. Despite the tension of the expedition and the mysterious deaths, she was now in her element. After all this time and planning and hoping and worrying, after the years of administrative hoops and red tape, she was on her way to the seafloor of an alien world in search of her passion—active volcanoes. These, of course, would be quite different from the standard fare she'd seen off the coasts of Oregon, Italy, Iceland and Antarctica. Pressures at those deep-sea volcanoes were much higher than here; Europa's gravity held its waters with a gentler hand. But Hadley supposed that the plumes might be just as scalding as the 200° C lava-heated plumes of Earth's ocean floor. She wondered if the alien hydrothermal springs carried the same sulfur-bearing minerals, perhaps mineralizing and growing exotic 30-meter chimneys like their terrestrial counterparts. Would they belch the same black clouds that gave rise to the name "black smokers" back home? Her thoughts wandered to the vents in the depths of the Pacific, or along the Atlantic ridge. There, 5 kilometers below the weather and sunlight, unaffected by the warming climate, they constituted a world unto their own. And like them, Europa's seafloor world was cut off from the elements above, sheltered from the vacuum and radiation. Like Earth's vents, would these alien chemistries engender those bacterial life forms that Dakota and Amanda so badly yearned for?

"We've got another couple hours," Cann called back. "You can always take a nap. Not much to see for a while."

Hadley took him up on the offer.

_*_

Brushing the sleep from her eyes, Hadley Nobile stared at her reflection in the porthole. Her face told many stories. She had the eyes, the crow's feet, the aquiline nose and the strong jawline of Uncle Nobile, but the resemblance went much farther than skin-deep. The same fires of exploration burned within her chest, the same determined quest to understand, to push the boundaries, to bring light where there had only been darkness.

The darkness outside served to reinforce the thought, but now, a faint flow emanated from below, a glow native to this eternally gloomy abyss.

"See it?" Cann called out. "Below, at about 11 o'clock."

"Got it," Hadley said, reining in her excitement. The ember glow of simmering lava told her they were getting close. Cann turned the sub toward the light, letting it descend on its own. Currents began to sway the ship. As Hadley felt the cabin rock back and forth, she tried to train her instruments in the direction of the approaching vents. Suddenly, the submersible lurched to the starboard.

"Whoa, Captain," she said. "Can you give my petite science gizmos a little smoother ride?"

Cann struggled with the joystick. "We've got some…ah…cross currents…and something else…"

"Something else? Like what else?"

"It's that idiotic engine again. It's overheating and binding up a bit. I'm going to let us glide for a minute, give it a chance to cool down."

Hadley spread her hand against the glass in frustration. She said nothing as the vent, which was beginning to show itself, drifted beyond her view. In the distance, she could see other plumes shimmering through the depths.

She leaned into the port on the opposite side. The waters there were radiant glowing now, orange fog billowing up. Beneath, the vents came into view again, and this time, they had resolved into a chain of lustrous yellow pearls embedded in a ruby snake wandering along the ridge. From the ridge, chimneys rose in fairy-tale structures.

"I'm going to try it again. Think we're okay for a while."

"I'm in your hands, Jeff," she encouraged. The sub descended into a forest of magma towers. Most were silhouetted against the orange glow beyond. She flicked on the external floods.

"Wow," Cann said.

"Wow is right." The chimneys took on the form of castle turrets with buttresses along their sides enough to make any gothic cathedral jealous. Other forms crowned the towers: cornice, cupola, tourell. Fluting embossed the flanks of many. Clouds of white detritus billowed across the field of view, then disappeared as quickly as they had arrived. On the far side of the grouping, a gorge wandered out from between two pillars.

"Can you get us down there? Down in that little chasm?"

"Outer temps are rising fast, but I'll bet we can get in closer before we have to leave."

"Just in case, can you hold steady for a moment while I take samples?"

"You bet."

"Already got lots of plume readings, but I want to get a rock or two, scrape some of that yellow stuff off the side."

The submarine's claw grabbed a small stone with the dexterity of a surgeon, depositing it into a bin. The video feed labeled the sample. Hadley took another, and another. She deployed a second device and scraped the surface, sealing the sample for later analysis.

"Okay, skipper, let's dive!"

Cann dropped the sub into the rift. The golden glow dimmed to near-darkness. Hadley could see tendrils of glowing orange marking the cracks in pillow lava. The black humps grew from the ground like biscuits in an oven. Not a bad analogy, Hadley thought. She grabbed another sample.

"I'm losing that engine again. Gotta take her up a bit."

"No prob. I'll just do some more imaging."

Cann shook his head with a smile. "It's Christmas morning for Hadley, with all the rocks and sludge she can unwrap. Glad it makes you happy."

"You have something against sludge?"

Before he could answer, the little sub jolted off its smooth course. "That's a pretty respectable current," Cann said, his voice clearly controlled.

Hadley knew it was more than a casual eddy. The thermal currents here would sear a human to cinder in moments. The sub continued to turn in the stream, lurching to one side and then the other. Hadley put her hand against the ceiling reflexively.

"Oops," Cann said.

"Don't say 'oops' unless you really mean it."

"Oh, I do. I think I've lost that engine. Protocol says we gotta go back up while I can still control this thing." It pitched again. "Moderately, at least."

As the vehicle began its long ascent, Hadley whispered, "Damn." She took in a long breath and watched out her window. "That's okay Jeff," she said aloud, "You got me nice and close, and I've got a quarter ton of stuff to carry back. Good job."

III

Hadley toweled her hair off as she sat down beside Gibson. She wiggled a pinky in her ear, trying to get the shower water out. Gibson hunched over his screen. He caught his breath, his face lit by the monitor. He smiled and turned to Hadley, pressing his hands together as if in prayer. "Want to see some more?"

"What do you mean?"

"More eruptions." He waved toward his monitor. "It appears that our geyser activity has finally arrived."

Hadley rocketed to her feet. "Yippee cay-yay, pardner. Let's tell the others. As soon as Jeff's got our little submarine powered off, we've got some traveling to do."

"We do," he said, but there was hesitation in his voice.

"What? After all this prep and planning and travel, is something giving you qualms?"

"Hadley, we've lost some people. Not to put too fine a point on it, there's a murderer among us and we're no closer to finding out who it is."

She shoved her hand through the air, pointing toward the window. "And there's a once-in-a-lifetime experience waiting for research. Ted and Orri and Joel wouldn't have wanted us to stop now. Let's make their deaths count for something. We owe it to humanity and to our team to play this out."

"I'm thinking more locally. Don't we owe it to our team to let them in on it?"

"How can you even say that? You can't equate the two. They are on entirely different scales. If we tell everyone, all they'll do is get stressed out and make mistakes and do stupid things to get someone killed."

"Someone has gotten killed, Had. We need to think of colleagues before we go off thinking about some abstract concept like future generations of researchers."

She shook her head slowly. "I can't believe I'm hearing this. I thought we were on the same page about all this."

"Believe it."

"Well, at the end of the day, I'm in charge of this escapade."

Gibson's clenched jaw betrayed a new understanding. "That guy back at Taliesin was wrong. Dakota's not Ahab. You are. And those geysers are your great white quest."

"Perhaps so," she said, her voice even.

"Just be careful of that harpoon cord."

Dakota passed by the doorway, not changing her stride.

_*_

"So you're *all* leaving?"

"Don't sound so surprised! I need them out there. You won't have a lot of help here for the next few days, so I just don't know that another swim for you is wise." Hadley scowled at Jeff Cann. In this cramped room, the submersible comm tech looked like a professional wrestler, a mountain of muscle. But he stood before Hadley with timidity written all over him. He held his shoulders tightly. He made eye contact with her shoes. Supplication came to mind. "Hadley, if we get this thing turned around, if I can fix our little problem— and it *is* a *little* problem—we could get another valuable dive in before you all are back. There's lots we could do in three or four days. Besides, you know Dakota is dying to get down to those hydrothermal vents."

"But I'm responsible."

"So delegate!"

She admired his fervor, and the expedition was on a deadline. While the guidelines called for the team to be out in the heightened radiation—far to the south of Taliesin—for no more than three weeks, that was at normal background levels. Io's volcanoes had suddenly become exceptionally energetic, pumping up the electromagnetic environs of Jupiter to near-record levels. And while those geysers were in the "hot zone" of increased radiation, even the relatively safe dome itself was vulnerable to higher levels than Taliesin was. At this rate they would need to leave for Taliesin shortly after the geyser team returned. And he was right: there were a lot of things that the home team could accomplish.

"Let me think about it."

Cann nodded, ground his teeth, and smiled as he left. It was a tight smile.

_*_

So Hadley thought about it. She thought about the crew who would come with her to the geysers, and the crew left behind. Her geyser team was the best she could ask for: geologists Gibson and Felicia Tanaka, and Tefia Santana.

And at home base, Sterling Ewing-Rhys, Jeff Cann, Amanda and Aaron Grant, and Dakota Barnes. Heavy on engineers and biologists. And while the entire biology contingent would stay behind, the geyser team would bring plenty of samples home. She wished she had another medical person to monitor the dive—Aaron's background was weak on radiation technology—but as with any fieldwork, they were stretched thin.

And she thought of those who had come before. General Umberto Nobile and his forlorn crew. And more recently and more to the point, David Culpepper's lost expedition, with their leavings scattered across the landscape just to the southeast. Was it too risky? Wasn't exploration about risk? Wasn't life? She was responsible for these people, as she had been, in a lesser way, for those on the lost expedition. No one seemed to know about her connection, and that was fine with her. But with responsibility, and with history, came consequences. She could not endanger anyone in the name of science.

She could sense Uncle Nobile frowning. Perhaps it was an expression of disapproval for her even considering the risky move. Or perhaps he was uncomfortable with her hesitation in the face of scientific opportunity, of adventure.

She remembered Ted's ripples in the pond of life. What was she going to leave to the future? The knowledge gained from a critical extra dive, or ignorance? Light or darkness? Graffiti or masterpiece?

She drew in a long, deep breath. Her hand shook as she keyed her wristcomm. "Jeff? You're on. You can take your dive while we're out."

10

Southern Realms

I

This was it—the culmination of a generation of scientists' work, the end game for an expedition six years in the planning. True, there were many reasons for being here: geology, macro and microbiology, studies of the alien chemistries in the ice. But for Hadley Nobile, the geysers of Europa were one of three primary reasons to be here. The second was the volcanoes undersea. Check. The third, the unofficial reason…

She wouldn't think about the third just yet.

Hadley surveyed the landscape to the south. There was something supernaturally clear about the view. Every edge was crisp, every subtlety of color cranked up in intensity. The black of the sky vibrated against the blaring white horizon.

There's nothing like a little peril to focus your mind.

Unlike the first expedition, the fast gliders had suffered no ill effects from the radiation. But they were headed into the same toxic world of cryogenic temperatures, vacuum, deadly electromagnetic forces. Europa was not a world to be taken lightly. And the time was ticking away. Their bodies were piling up the rads, and the medical rules dictated that they could be allowed only so much. Perhaps that was a good thing. Perhaps they would have to get back to Taliesin before the murderer could act again. Then, she would arrange for separate ships from base to take some of the crew. Split them up. Divide and conquer. She and Gibson would hold back for a later flight.

Gibson. Not many things in the world stayed the same, but he was one. Gibson was more than her rudder; he served as a sort of anchor to sanity.

© Springer International Publishing Switzerland 2017
M. Carroll, *Europa's Lost Expedition*, Science and Fiction,
DOI 10.1007/978-3-319-43159-8_10

When the danger and death and, yes, *murders* ate away at her calm foundation, she simply looked to him as a reminder that life still held good things and good people. She couldn't think of anyone she would rather be with to witness, for the first time up close, the fountains of Europa. But now they were on different pages, it seemed. He thought she wasn't being prudent. Perhaps he was right. But she could really use his support just now, and for the first time, she didn't think she had it.

—*—

"Hey Gibson, I have something for you." Dakota entered his room as he finished stocking his pack.

"Oh?"

"Good luck charm." She handed him a tiny hammer, a little rubber model from some obscure set of toys. "Got it at Taliesin, in that little shop."

"Thanks, Dakota. This will come in handy."

She punched him in the shoulder. "It's so you won't forget our, shall we say, future options?" Her bluster faded as a blush tinted her face.

He kissed her on the forehead, stuffed it in his shirt pocket, and headed for the airlock to suit up.

Dakota wondered what had just happened. Clearly, there was tension between Hadley and Gibson, but they didn't seem to be romantically inclined. And he kissed her on the forehead? Her forehead! As if he was trying to extricate himself. The man was complex. Or maybe just confused. With all the men she had known in life, she could certainly believe that.

Dakota made her way to the comms station, where she watched out the porthole as the glider slowly turned away from the dome. The lights of the markers at the edge of the base's magnetospheric field lines glowed red. They blinked off as Jeff Cann shut down the shield. The vehicle paused.

"What are they waiting for?" she asked.

"Permission to go." He squinted through the window as if trying to see the little bolts on the glider, then turned to Dakota. "See, even though this thing is built to protect us from all that nasty energy coming down, radiation actually gets trapped within it. It builds up along the shield's field lines, and when we shut it down we have to give it a chance to dissipate for a few moments before they can go out. Hang on."

He keyed his microphone. "Okay guys, all clear. Godspeed."

"Roger that," came Hadley's voice over the intercom. The glider slid past the row of cones. The lights flared again as Cann brought the shield back online. The vehicle remained visible for a good five minutes, beyond the protection of home base, bathing in radiation. It snaked in and out of hills and depressions, finally dipping behind a reddish mound. With its departure, Dakota felt oddly empty.

Cann put an arm around her shoulder, gently. "You okay?"

"You bet," Dakota said a little too quickly. But she felt comfort from the man's arm across her back. "I'll buy you a cuppa, and you can tell me all about your, ah, shield."

II

Tefia had watched, longingly, as the base disappeared beyond the low ridge just beyond the markers. Now, they were on their own.

A million-year drizzle of micrometeoroids had scoured Europa's ice plains into a frosted-glass patina. Here and there, the cliff faces had splintered off, leaving an alarming glacial blue stripe shining out. The terrain became jumbled and treacherous, making driving exhausting. The glider skimmed just above the surface, but abrupt rises or boulders had a tendency to bark the undercarriage. Hadley filled in for Gibson after two hours, and Tefia took a turn in another hour.

With Felicia now in the driver's seat, the glider slid up a steep ascent, the rim of a small chaos collapse zone. As it crested the hill, the black sky began to glow in a bluish haze.

"There they are!" Felicia called, pointing out the windshield. At the horizon, curtains of mist undulated, fanning out as they rose into the sky.

"Look at the base," Hadley said. "They don't look like single-point eruptive vents. They're more like rifts."

"Yes," Gibson said, staring. "More Enceladus than Etna. But these are a whole lot fainter than anything I saw on Enceladus."

"They remind me of the aurora," Felicia said, letting the glider coast to a stop.

The sheet of material wavered in great curtains, then abruptly brightened as a denser wave of particles migrated up the plume. It trailed off, drifting down toward the ground in a direct course, without curving. Hadley had to remind herself that the plumes were operating in a vacuum; they betrayed no sign of billowing or drifting across the landscape on a breeze. Nothing disturbed the ballistic flight of the particles. Another stream of material appeared to the side, adding to the long line of ghostly eruptions.

Tefia felt like a tourist gazing at a new and unfamiliar vacation destination. The others had work to do. She had to watch them. She glanced down at her readouts. The radiation counter had plateaued over the past half hour, but now it was making its way back up the scale.

Hadley looked at Felicia, still behind the joystick. "Well, shall we?"

"Yes, of course." Felicia brought the vehicle over the hill and around a rough outcrop. As she did, the Sun moved behind the plumes, scattering beautiful rainbows of color through the ice crystals. Everybody but Felicia was taking photos.

"Let's see if we can get over to that flat spot just east of them."

Felicia turned the glider toward a steep talus slope.

"Your other east," Hadley said. Felicia cranked the joystick back. The edge of the glider bit into the scree. The entire vehicle rocked, righted itself, and stabilized.

Felicia let out a breath and coaxed the vehicle back the other way. An alarm sounded. She held up a hand. "What did I do?"

Gibson leaned over the console. "Nothing you did. Radiation alert."

"Damn," Tefia snapped.

"No worries," Hadley said calmly. "We'll set up camp at a safe distance and start chopping ice blocks for the roof. That, along with the cab's own insulation, should give us a nice little shelter, right Tef?"

Tefia nodded tentatively, casting a wary eye at the monitor.

Felicia threaded the glider through crags and ice boulders, finally settling on a flattened depression. The geysers loomed just beyond a disorderly rise a dozen meters from them. Its surface sparkled in the sunlight, glimmering with freshly deposited frost. She hit her laser rangefinder. It wavered as it tried to read the ethereal undulations of plume material.

"Looks like we're at about a 60-meter range. Too close?"

Gibson interjected, "You kidding? Hadley'd rather be right inside those things, but it's close enough for me."

Tefia couldn't tell whether Gibson's remark had been a good-natured tease or biting sarcasm. He and Hadley had clearly been tense lately. Then again, exploration was a tension-building exercise. Hadley said, "This will be just fine, Felicia. Great job."

Felicia let the craft settle to the ground. She and Hadley unstrapped. Gibson and Tefia had been standing behind their seats anyway, watching out the cockpit windshield. Not very safe, Tefia reflected, but perfectly understandable under the circumstances.

"Everybody suit up. Time for a reconnaissance. Tefia, you know the drill."

She did. Tefia and Felicia were to set to work at once cutting blocks of ice for a makeshift radiation storm shelter on the roof of the glider's cab. Hadley and Gibson entered the small lock, preparing to set up a remote station near the geysers, in case anything went wrong and they needed to make a hasty exit, stage left.

As Gibson and Hadley pumped the airlock, Tefia radioed, "Remember: I'll be feeding you regular reports on our incoming radiation levels; I'll be watching the Io reports so you don't have to. If Io gets excited and things get bad, you come home and we go into our glider's little makeshift storm shelter. Doctor's orders."

"Got it," Hadley said. "Airlock secured."

Tefia looked out the side window. Hadley and Gibson were walking toward the back of the glider, where the equipment bay was. The lock indicator read full pressure once again. Tefia and Felicia stepped in and sealed the hatch. Felicia wore a nervous look.

"Nothing to worry about," Tefia said. "I do this all the time."

"You're a lousy liar, Doc. I like that."

They both grinned as the air drained from the lock.

III

"Beautiful," Gibson puffed. "Can't wait to see … what's over the hill. Man, for an eighth of a g, I'm bushed."

Hadley looked down at him from slightly above. "Hey, what doesn't kill you makes you … " Her breath ran out.

"I know, I know. Let's—" He held up his hand. Hadley nodded and stopped to rest.

They stood in the shadow of the steep hill, a dark slope of talus fanning out below them. As they faced away from the hill, Hadley realized that the sky was no longer the velvety black so typical of Europa. The stars, visible now in the shadows, floated in a deep metallic blue mist. On the horizon, directly away from the hidden Sun's light, a rainbow of color glowed at the anti-solar point, undulating in the wafting haze.

"Do you feel it?" Hadley asked. "Under our feet?"

"Feels like rumbling. Like thunder."

"Let's go!" she said. Through his helmet, Hadley could see the mixture of feelings on Gibson's face—disappointment in not taking more time to catch his breath, eagerness to see the geysers they had come so far to witness. The latter clearly overtook him, as he began with renewed vigor. It was overtaking her, too. The geysers were blowing away the dark clouds of homicide and intrigue.

Hadley crested the hill. Sunlight streamed onto her face, momentarily blinding her. As her eyes adjusted, a deep fissure sprawled before her, snaking its way to the horizon in great parallel bastions of glowing ice. The blizzards of erupting material had transformed the landscape into a wonderland.

"Dear Lord in Heaven," Gibson said.

Hadley couldn't find any words. She began to hyperventilate. She forced her breathing to slow.

The slope facing the geysers gave the appearance of a pine forest bleached of all color but faint blue and rust. Great spikes rose from white mounds, clawing toward the sky like the talons of a titanic monster. Their points aligned in the direction of the geyser activity, like iron filings on a magnet. A dusting of glittering powder covered the tops of all the outcrops in a jewel-like blanket. The sparkling powder banked along ice-rocks and piled in drifts like Sahara dunes. For a moment, she had a vision of Ted sitting on the slope with his sketchbook, taking it all in.

"This a good spot to set up?" Gibson whispered, awed into quietude by the scene.

"Maybe we should get closer."

"The whole idea of a contingency hike is to set something up in case we can't get closer, in case we have to leave. You know how volcanoes can be."

He was right, of course. Hadley had watched lava flows change course as abruptly as a snake slithering through a garden, taking buildings and people with them. She had seen fissures open up across stable ground far from volcanic vents, providing nasty surprises for her teammates.

"Here is good," she said. "But then let's go down a ways." She knew that the geologist in Gibson couldn't refuse the idea.

They set to work quickly, setting up the remote science monitoring station atop the hill's crest, in direct line with the glider. A smaller, omnidirectional antenna could send bursts of information to Europa's satellite constellation when one of its members passed near enough for communication.

Gibson flipped a small switch on the side. "Data is streaming."

"Excellent. Shall we?"

Hadley felt like a kid bounding down a snowy hill, wending her way joyously through the forest of ice pillars. At the bottom of the slope, the two crossed a hummocky ice rink to a rise on the other side. They were close to one of the plume vents. Ice crystals stuck to their helmets, texturing their view with the quality of confectioner's sugar. Hadley paused to take in the towering plumes, draping mantles of light against the dark cosmos. The closest plume was less fire hose and more diaphanous filaments.

Suddenly, the ground seemed to tilt away, dumping both of the explorers off their feet. The solid ground became an undulating, living thing, flexing muscle, pulling sinew, with the mind of a predator. Hadley felt, more than heard, a sharp crack. Parallel to the geyser ridge, a dark zigzag opened up, etching a black lightning bolt upon the sparkling surface. As she struggled to her feet, the new fissure continued to widen all along the bottom of the hill they had descended moments earlier.

"Cut off," Gibson said unhelpfully. Gravel from the hill began to slide into the crack. Hadley looked up at the little science station, sitting on its tripod, and wondered if it would survive the eruption without tipping over.

Surprisingly, Gibson was snapping photos of the geysers and the fracture. He said, "I've seen something like this on Enceladus. At least the leftovers of it. Maybe if we go left we can find a way back."

"That crack can't go on forever," she said.

"You hope."

—*—

"It's a whole lot cheaper than radiation-hardened blanketing," Felicia told Tefia.

"Yeah, and you don't have to bring it with you." Tefia hefted another block of ice up to Felicia, who had stationed herself on the glider's side ladder.

Felicia grunted, slid the brick into place atop the glider's reinforced cab, and said, "That ought to do it." She glanced back toward the rear. From her vantage point, she could just see the inflatable habitat extending out the back end of the glider. Thanks to the balloon-like structure, they now had sleeping quarters.

"Let's check inside," Tefia said. They climbed aboard, pumped up the lock and unsealed the hatch. Tefia removed her helmet. Her bangs stuck to the perspiration on her forehead. She took readings inside the airlock, in the main cabin, and in the sheltered cockpit. Radiation levels within the cockpit dropped dramatically.

"There you have it," Tefia said. "Instant storm shelter."

Felicia shoved her nose up against the window. "Where could they be? I thought a contingency trek was supposed to be quick and dirty."

"It is. No idea."

IV

Fine ice crystals covered most of Hadley's faceplate. As she tried to scrape her visor off, she felt the thunder beneath her boots. The heaving ground reminded her of a dangerous eruption she had experienced—all too closely—in Indonesia, an experience she did not want to repeat.

"Here, let me get that." Gibson scratched away at her helmet visor until she could see out. With each swipe of his hand, her heads-up display glowed like a surrealist's green masterwork.

Hadley took a turn clearing his, and then said, "Let's get outta here. This way."

She followed a shelf along the contour of the fissure. The hill with the automated science station remained intact, for now. The shelf met the floor of the valley. The ice dust piled up in hillocks and crests, crippling their strides and slowing progress. But she could see the crack up ahead, and it was petering out.

"I think we can cross down there." She pointed. As she did, another loud pop resounded through the ground. Behind Gibson, the ground simply fell away. "Run!"

He bounded toward her as layer upon layer of ice fell into the widening abyss. Streams of water blew into the sky, dropping a pummeling hail of ice across the landscape. The patter of hailstones was deafening. Gibson was trying to tell her something, but she couldn't make it out. All she could do was make her way back up the side of the hill, away from the maelstrom.

"Around here!" Gibson called, gesturing wildly. He was headed for a saddle between two hills. Hadley followed quickly. They crested the rise and turned to look back.

Now beyond range of the hail, Hadley wilted in relief. "We made it."

The little science station stood guard on the other hill, stark against the billowing sky. Suddenly, it tilted. The entire face of the mound dissolved before their eyes, sliding into the black fissure in a cloud of ice crystals.

"So much for contingency." Hadley said.

"How about redundancy?" Gibson asked. He pulled a smaller version of the lost science monitor from his backpack.

"Nice thinking."

"I'm setting it up right here. Respectable view of the geysers, nice stable spot of land."

"Less dramatic, but much safer," she agreed.

_*_

Felicia always hated to be the bearer of bad news. She was a people-pleaser, and this news would not be pleasing anybody. She tapped her comms monitor again and shook her head at Tefia.

The voices of Hadley and Gibson finally broke through the static.

"Glider One, do you read?" Hadley's voice cracked in the headset.

"Affirmative, just barely," Felicia replied, relieved. She could see them now, cresting a rise to the east, making a beeline for the glider. Their suits glistened with frost.

As the airlock settled and Hadley cracked it open, Tefia spread her feet, jammed her hands on to her hips, and spoke in a motherly voice. "Okay, kids, where have you been? We've been worried sick!"

Hadley pulled one boot off. "We had a geyser adventure. Are you getting data?"

Felicia tried not to look sheepish. "We've had some … issues."

Hadley frowned. "As in?"

"As in, the comms seem to have gone to hell," Tefia said.

"Could it be radiation? Magnetospheric interference?" Even as Hadley offered the question, Felicia shook her head.

"Io's quieted down nicely. Ambient rads are fair to midllin'. It's got to be something mechanical."

"We'll have to check it out." Hadley sounded determined.

Tefia held up her hand. "For the good of our little crew, it can wait until tomorrow morning. You guys had a long day, and so did we. Now that camp is all set up and we have some room, let's get some chow and some shuteye."

"Sounds great to me," Gibson said, tossing his gloves onto the airlock shelf.

"Shame," Hadley said. "The team back at the dome should do their third dive tonight. Wish we could hear it."

V

Gibson had the windows shuttered and the glider's red lights on as a reminder that it was nighttime, despite the blaring whiteness outside the panes. Tefia and Felicia had retreated to their sleep chambers in the rear. Gibson and Hadley sat in front of a screen, which seemed to be the only way they communicated of late. The fragmented diary of the lost expedition glowed across the monitor.

"Remember," Gibson told her. "Those guys weren't interested in geyser activity. It was beyond their scope. Except for a quick deployment of the remote science station to the east, it was strictly drill and dive for them."

"Yes, at that point nobody knew about the periodicity of the geysers," Hadley mused quietly, gazing at Gibson's monitor. She waved toward it. "So let me have the rest."

"Right." He tapped the screen and text began to scroll. "So, to review, at this point in the log they have lost just about every mode of transportation—"

"Wait, let's take your overview back to the departure. I need a refresher."

"No surprise. It's complicated. So they depart from Taliesin February 6. They arrive at the south dome about six days later, at which time Donnie and Genevive are dispatched to set up Remote Station B to the east."

"Well into the hot zone," Hadley interjected.

"Yes, high radiation there, so they don't stay long. They're back by the 17th. By then, the bore hole has broken through the ice, all prepared for the sub, and Alison has had her breakdown and somehow killed Peter Kaminsky. A day later, they lose Serge Montano in the submersible. Between February 20th to 22nd, Primary Rover 1 fails."

"And Alison dies on the 22nd?"

"Correct, the 22nd. And because of her death, they now have room to return in Primary Rover 2, but the thing breaks down on the way because they're pulling some kind of sledge."

"What could they have wanted to pull along with them?"

Gibson swiveled to look at Hadley. "I'm thinking they were trying to bring the bodies back with them."

"Not the best decision in a survival situation."

"Twenty-twenty hindsight. So they essentially pitch camp where they are, somewhere on the way back to Taliesin, but probably still a bit closer to the dome. With Primary 2 now partially crippled, they cannibalize parts from Primary 1, which must have been fairly nearby, but they're missing some critical component that they think is at the eastern cache, so Donnie Ramirez and Dave Culpepper set out on the minirover to get there."

"And Culpepper totals the minirover."

"Right. If it wasn't so tragic, I'd call it a comedy of errors. Primary 2 makes it to them, somehow, and takes everybody back to the camp, where Genevive dies. Mendelson commits suicide by radiation, which leaves Donnie and Culpepper to set out for the southern dome on the other minirover."

"I thought their secondary minirover was a one-person deal."

"That's what this next bit is about. Ready?"

I cannot believe the ingenuity of Donnie Ramirez. He took what was left of our miniro … like some stunt driver's motorbike.

"So now they've got the minirover back on line," Hadley commented. "They must have modified the other mini as a two-wheeler and got it working again."

"Maybe. But the last entry is crystal clear, as cogent as anyone could be." He pointed to the screen.

He made a cart to pull me with the modified mini. Isn't that thoughtful? But I'm going to turn off my suit before I'm a burden to him. No way he can drive on a glorified motorcycle all those kilometers pulling extra weight.

The words scattered goosebumps along her arms. Had Culpepper abandoned Donnie, walking away like an Inuit's death trek?

She frowned. "But Donnie's body should have been found inside or near the dome, because that's where they found the diary, and we know Culpepper didn't make it down there."

"Yeah, but what happened to the scooter he was driving? And here's another weird thing: the footage from the recovery team shows some scratches on the dome wall. Donnie had survived to make four-day's-worth of hatch marks in the metal. Four long, lonely days. That's how much air he had left, unless he found another stash at the dome. And there's also a mysterious drawing, sort of a diagram—its like a half circle of five points with its diameter on a horizontal line so that the circle points up like a sideways letter 'D.'"

"Yes, I saw that. And another small circle above it and an 'x' below."

"That's not all," Gibson said. "Donnie apparently made one last entry in the diary. It's fragmented, but I recognize the statement. It's a quote of Camus, the French philosopher: 'Without culture, and the relative freedom it implies, society, even when perfect, is but a jungle. This is why any authentic creation is a gift to the future.'"

"Curious," Hadley said. "Was he referring to some creation they made?"

"Or maybe to the creation of the outpost?" Gibson offered.

"Or the creation of knowledge, or something more concrete?"

"All good questions. So where does that leave everybody?" Gibson asked.

Hadley went over the numbers again, trying to solidify them in her mind. "The recovery team found Genevive's body in the main rover, as if it had been stowed carefully after she died. Mendelson was found at a distance in orbital images, where he ended up after he took his last long walk. The bodies of Peter and Alison remained at the dome in cold storage, so to speak. Dave Culpepper was never found, but I'll bet he's somewhere between the site of the Primary Rover and the dome."

"Or maybe closer to the Remote Science Station B?" Gibson offered.

"Maybe. The diary ended up with Donnie, who left it at the dome and disappeared. Why? And Serge was never found, obviously."

"He had a burial at sea," Gibson intoned.

"But the key may be out at Science Station B. What were they doing out there, and why was it so controversial to some of the team? Clearly, Mendelson had some kind of issue with the activities out there."

She turned her attention to the view out the small porthole. "Odd not to see Jupiter in the sky."

"Just a couple hundred klicks or so over there if you get lonely for it. Just a short dri..." The word faded away before he finished it.

"Drive?" Hadley said for him.

"Yes," Gibson said, slowly, his eyes focused on another time and place. "Yes, just a short drive from here. And shorter from … "

"From where? What are you talking about?"

He shook his head, as if trying to awaken from sleep. "A shorter drive from where the first expedition was. You could see Jupiter from their eastern science station. I think I just solved a little historical mystery. But I'll need to check some things."

"What, in your spare time?"

"Make up your mind," he snapped. "Should I be working on this or shouldn't I?"

Hadley's words were measured. "It would help us reconstruct all that went on."

"Look, Had, out at that abandoned cache, there's something there. Something no one's ever noticed. Nobody did because it's made of ice, but I think I've figured it out, from a couple cryptic fragments in the expedition records. And from Donnie Ramirez's doodle on the dome wall. I know it's still there. I think we should go see it. We'll need to go out there, on site, to figure out those last days of the expedition anyway."

Hadley rubbed the back of her neck. "If it will help us figure out what happened back then, maybe so. But is it worth the risk, going that far east to be sure?"

"I would say so. The earlier search parties didn't go that far out of their way. Back then it was too difficult. But whatever's left out there might show us what they had to work with, and that may be where Donnie and Culpepper ended up."

"Agreed," Hadley said.

"And I've decrypted another little snippet that might have a bearing on all this. Where is it?" He scrolled down ribbons of text and fragments of gibberish, finally coming to rest on a short couplet. "Here it is."

While we're waiting for the cavalry, construction progresses on our little project. They say Primary 2 will be able to fetch us. Hope so. This is a lonely place to end it all … stranded. I suppose at this point it won't hurt to spend a little time on something like this, a monument for the next explorers, God bless the poor bastards. Mendelson called us "little kids at play," but at the end of the day, this may be the only legacy we leave.

"What could that mean?" Hadley's voice rang with frustration.

"I have a hunch, but we'll need to see."

Hadley pondered the options for a moment. "And with our geyser observations finished and the cryovolcanic activity dormant again, I doubt that anyone will venture this far south for some time to come."

"Which makes it even more important that we get to that abandoned cache at Station B. The way that radiation and vacuum erode things out here, this may be the last time anyone has the chance to find out just what happened in the last days of the lost expedition."

"Yes, I suppose so."

"You sound disappointed."

The corners of her mouth twitched. "I guess I was hoping for some smoking gun, some specific reason that it all went south for these people. Radiation, hardware failure, DSN, it's just not very tidy."

"No, it's not. A cascade of events is less satisfying than being able to say, 'They failed because of X.' But I'm sure it was all dramatic enough for them at the time." He frowned.

"I know that look," she said. "Something's distracting you, and I'll bet it's not about that lost expedition."

He nodded, quieting his voice. "I was just wondering about something else, something pertaining to our *other* problem. Aside from the fact that we're running out of suspects, I wonder what kind of outside help somebody might be getting."

"Don't tell me you're beginning to believe in conspiracies."

Gibson didn't smile. "If I'm right, we have the mother of all conspiracies right under our noses. But I'm going to need a little time to dig into history."

"Ancient history?"

"War history. Think about it. Sterling was involved in the same arena as the greatest war criminals were, the region of the camps."

"And the Seven Sisters."

"Yes. Now, Amanda and Aaron are eccentric, I'll grant you that. But Sterling has baggage. My money's on him, and I think we need to be careful of him."

Hadley felt as if the chill of Europa had just tumbled down her spine. "A conspiracy has to involve more than one person."

Gibson studied her. "At least two. That bothers me, too. But until we find out about Sterling, we won't know about who else is involved."

"So now what?" she whispered.

"I suppose we just watch them like a hawk—all of them, including Dakota—and if anybody does something weird, we confront them in front of the whole group."

"Why wait?"

"No evidence. They could laugh it off."

Hadley leaned forward. "But it might stay their hand."

"I doubt it. They're already in way too far to stop now. There's really only one solution. We've got to put it out in the open, to let everybody know that somebody in the group is guilty of multiple murders."

"Absolutely not!" she barked. "That will only serve to slow us down. I need my science team to be focused on science. If everybody's glancing behind their backs, nobody's going to be getting the last bits of data from here, and that's going to be some precious data."

"Life is precious," Gibson said. His gaze was level, his eyes intense.

"Of course it is. Don't make me out to be the villain here."

"We can arrange duties so that everybody is safe. All it takes is three people doing each task. With three, no one can be 'knocked off.'"

"The sub only takes two passengers."

Gibson sounded a note of frustration. "Hadley, you know what I'm saying. I just don't get you. First, you hide the fact that your father was involved in the war."

"I didn't hide anything. I just didn't think it was appropriate to drag it into the conversation at the time. Just because you know all my secrets doesn't mean that everybody needs to."

"And now you pull this. I say damn the science. Keeping this thing secret is not going to help anybody's health."

"I'm not pulling anything. But I *will* pull rank if I have to."

"Rank? Seriously? With all that's at stake here, just figure it out. What are your priorities? What's best for the team?"

"I'm in no mood for a lecture. We'll have everybody work in teams of three, fine. We won't say why. We'll make it a ... a guideline."

"A rule," Gibson shot back.

"Fine." She said the word as he headed toward the sleeping hab at the rear. Gibson was her closest ally, and now he was furious. Just the frame of mind that a murderer could take advantage of. The dark thought seemed to shadow the pristine ice outside.

11

Mayday

I

"Third dive is officially under way," Jeff Cann said into his headset. Behind him, Dakota Barnes fidgeted with excitement.

"Settle down, Doctor Barnes," Cann scolded. "Save your energy for our upcoming adventure."

"Aye aye, skipper." She couldn't keep the mirth from her voice. Not very professional, but there it was. They were on an outing of the very best kind. At that moment, Dakota Barnes realized that this voyage, this expedition into the darkest waters of an alien moon, just might be the high point of her life. She smiled and took in a long, languishing draft of the cool cabin air. She stretched in the cramped quarters.

"You remind me of a cat," Cann called back.

"I promise I won't scratch your boat's beautiful interior."

"I like cats." He said it quietly, but she heard him nonetheless.

Outside, the darkness of the tunnel undulated with subtle layers in the dim light. The passing sedimentation reminded her of an express elevator she had once boarded in the Empire of Luxembourg's Steichen Tower, tallest building in the world. The speed had been dizzying.

Soon, the tunnel opened out into the vast expanse of Europa's alien ocean. Somewhere down in that darkness, Hadley's volcanoes were rumbling away. Dakota saw them as the best possibility for large-scale biota, but it was a long way down. She flipped a few switches, but the experiments in the bay had already activated. She reached up to her headset, tapped the screen, and said,

© Springer International Publishing Switzerland 2017
M. Carroll, *Europa's Lost Expedition*, Science and Fiction,
DOI 10.1007/978-3-319-43159-8_11

"Hey Amanda, are you using the instruments from up there? You already had your turn, girl." She heard only static.

"Amanda, do you read?"

More static.

"Amanda?"

The earpiece crackled. "Affirmative, Dakota. Nope, whatever it was, I didn't do it. Aaron says we're suffering a radiation spike. Maybe that's it?"

Dakota could see Cann nodding. "It's screwing with my readings even down here," he called back.

Dakota tapped her earpiece. "Hope all three of you are together, watching each other's back. It's important, you know." She didn't try to hide the sarcasm in her voice.

"Roger that," came Sterling's comment. "One big happy family here."

Dakota redirected the instruments, playing the suite of equipment like a concert pianist.

"More organics," she said. "Complex ones."

"But no blue whales?" Cann quipped.

"No whales of any kind," she said despondently. "I'd take paisley ones at this point. Captain Ahab, my ass."

"Let's take her deeper," Cann said. The submersible tipped forward into the darkness.

II

Gibson waddled over the rise, carrying an empty box, undoubtedly leftovers from his deployment of another remote station near a promising-looking vent. With the stations set up along the main fissure, all remotes were now deployed and operating. Hadley continued to help Tefia pull supplies from the glider's undercarriage.

Gibson gently grabbed her elbow, turning her to face him. "Hadley, we're in good shape. Every science station around here is sending its data to us and to Taliesin. It's time."

"For what?" She felt defensive.

He tensed his jaw, began to respond, and apparently thought better of it. He turned her toward the landscape beyond. "For this. To take it in. Not as a scientist on duty, but as a regular, run-of-the-mill human being. Not a human doing, a human *being*. Step up on this rock and just … *look!*"

She did. Ganymede hung very low in the sky, just gracing the distant mesas. The two explorers stood at the front of a natural corridor of frozen shelves and ridges, icy walls marching in stately aisles toward the horizon as neatly as if Frank Lloyd Wright had erected them there. To the left, one of the eruptions rose nearly overhead, arching to the right to meet the undulating curtains from the main fissure. The meeting of the two plumes in the sky gave the appearance of a great gothic arch, a sparkling cathedral of impossible proportions. Within the misty firmament, hints of Europa's greenish aurora played with the scattered colors of faint rainbows. Stars actually twinkled behind the flowing blizzards of water and CO_2. The low, Gregorian-chant rumbling in the ground added cadence to the cosmic basilica. She could easily imagine a choir of monks making their way down the central causeway, voices echoing against the cavernous nave. How she wished Ted could have seen it, this tribute to his Creator.

It could have been such a nice moment, but Gibson's directorial manor rubbed her the wrong way. He had won their argument, even though he was wrong. A little ignorance could be bliss in this case, and she would take blissful scientists over paranoid ones any day. Sending her colleagues around in triads was simply too far toward the cautious end of the spectrum. Three is a crowd, she thought. Even in research.

"That can't be right," Tefia's voice came over their headsets.

Hadley pulled herself away from the natural splendor, alarmed at Tefia's tone. "What is it?"

"I was just switching over to our return oxygen tanks. We seem to be … out. Can anyone here hold their breath for six hours?"

III

Hadley ducked out from under the glider and faced Gibson and Felicia. She keyed her mic to talk to the cockpit. "Any change, Tefia?"

"Still pegged on zero," she radioed.

Hadley stared at Gibson. "The emergency tanks are empty as well. Looks like one of the lines blew out." Through his visor, she could see that he looked as worried as Felicia did.

Gibson looked down, shuffled his feet, and then said, "Look, I can see us losing one of our return tanks. That's obviously why we carry three. But the emergency set is on a completely isolated system. Unless we've got some kind of cross-system leak, somebody drained it."

"Yeah, well, while you're marveling at our great bad luck, there's something else. When that little line blew, it damaged the cabling leading to comms.

We'll be lucky if the system will be able to hear us talking to ourselves, let alone communicating with anybody out there."

"This is bad," Felicia said. "This is really bad." She couldn't catch her breath. Her visor began to fog up. Hadley grabbed her by the shoulders, resting her helmet faceplate against Felicia's, looking into her eyes.

"The brainy engineer types among us just got this news. Let's give everyone a chance to process this little setback and see what we can all come up with. For now, let's you and me get inside and try to raise our colleagues at the dome. Good?"

Felicia nodded, her bobbing head just visible in the darkened helmet.

"I might have something that can help us," Gibson said. "I'll check out my little idea."

"There," Hadley told Felicia. "I told you it would just take time, right?"

As Felicia stepped up into the airlock, Hadley turned toward Gibson. He didn't look like a man with a solution. He looked like a scared puppy.

And the ghosts returned. Uncle Nobile was watching—she could feel him. And those who had been lost before, they laughed and cackled in the radio static in her ears. Instead of a chance at redemption, it was all happening again. A crippled rover. Deaths among the crew. A failed expedition.

In the moments that it took them to pump up the lock, Tefia had made them both coffee. She handed each of them steaming cups as they entered.

"You guys have been out a long time. This will help you focus."

"Thanks, Tef," Hadley said. "Actually, what I'd really like is a new high-gain antenna, but coffee will be good for now."

Hadley and Felicia made their way to the cockpit. Hadley put on a headset and tapped the screen. "Glider to base. Glider to base. Do you copy?"

She heard nothing in her earpiece but the banshee cry of Jupiter's radiation fields.

"I'll switch to X-band. Maybe that one works." She tapped her touchscreen again. "Glider to base. Come in Sterling, come in Amanda, come in Dakota. Glider to base. Do you copy?" She fought the panic that she could feel welling up within her own voice. "Glider to base, Glider to base. Do you read?"

She tapped another pad and looked at Felicia. "I'll see if we can send anything through our omnidirectional antenna. I know it's got low data rates, but it might still be working." She held her finger against her ear to listen more carefully. "Glider to dome. Glider to base. Pan-pan. Pan-pan. Panpan. Do you read?" Hadley realized that this was the first moment she had used the universal distress call 'pan-pan', something urgent, just a step below "mayday". Things must be getting to her.

Felicia turned up the volume. The rise and fall of Jupiter's deadly rain of electromagnetic particles hissed like waves on a beach. The sound of haunting shrieks betrayed the sing of Io's lethal flux tube, chaining the volcanic moon to its parent planet with that searing mix of charged energy. But in the midst of the noxious symphony, she heard no voices. She looked at Hadley. Hadley shrugged.

The lock cycled and Gibson entered. "I brought a nice little backup, a fuel cell."

"A fuel cell?" Tefia scowled. "Seems like we've got plenty of power. Last I checked, fuel cells require an O_2 tank to combine with hydrogen, right? I mean, isn't our problem a *shortage* of O_2?"

But Hadley smiled knowingly.

Gibson explained, "Before we left, I was thinking we might need the power from it, but I might be able to jury rig it to operate as the mirror image of the ones we use in space. Instead of it generating energy, we put energy into it. Like Hadley said, we've got plenty to spare. Then we use water ice—hydrogen and oxygen—in a reverse process, opposite to what it's intended for. The cell would no longer generate electricity. Instead, the thing would have two byproducts as it melts the ice. Hydrogen, and…"

"Oxygen," Tefia grinned. She began to shake her head slowly. "You are good."

"We just vent the hydrogen out and pump the O_2 into our tanks. The only problem is that it will be a slow process. We need to get the thing to operate right away if we're going to have enough air. I'll set it up to pump right into the secondary tank."

Felicia became animated. "We've already got ice blocks on the roof of the cab. We'll take them along in the external storage shelf and melt as we go."

Tefia turned toward the lock. "I'll bring down a couple now, and we can get it started."

"I'll help," Felicia said, reaching for her helmet.

"We'll need to drive fast," Hadley told Gibson.

"No problem. We're highly motivated."

12

Once More into the Fray

I

Hadley settled into the driver's seat. She took one last look out the windshield and spoke softly to Gibson. "I remember as a girl launching model rockets. I would look up at the sky, and no matter what my mind told me, that sky just looked like a big, smooth dome. Like a Robin's egg inside out. But then I'd fire off one of my little rockets and it would punch a hole right through that dome, make it three-dimensional, something deep and not a ceiling. Up and up and up, and you could finally see how big and deep the sky really was. And that's what this is like."

"Our geysers are punching a hole in your little black dome?"

She nodded, grinning. Sliding her sunglasses from the top of her head down onto her nose, she called over her shoulder. "You kids all ready to hit the road?"

"Ready when you are, bwana," Tefia shouted back. "We're just happily melting ice back here."

"Yes," added Felicia. "I'm too fond of breathing to give it up."

Gibson keyed his mic. "Glider to base. Sterling, do you read? Gibson to Sterling. Do you copy?" He glowered. "Nothing."

"Drained oxygen and no radio," Hadley whispered to him. "Somebody really wanted to give us a one-way trip."

"Good thing I'm so brilliant."

"A regular Boy Scout." She glanced back at Gibson's fuel cell, sprawled out on the floor of the crew compartment. A line fed directly from its side into a

© Springer International Publishing Switzerland 2017
M. Carroll, *Europa's Lost Expedition*, Science and Fiction,
DOI 10.1007/978-3-319-43159-8_12

jack on the wall, and from there into the spare oxygen tank below decks. The process was painfully slow, but it seemed to be working.

Turning back to the controls, Hadley carefully maneuvered the glider down the face of the scree and onto the plain spreading before them. The craft pirouetted a meter above the surface, then dipped slightly as Hadley urged it on. After a few minutes, when the glider was clear of the rugged terrain, she glanced back again. Both Tefia and Felicia had settled into their seats comfortably. Tefia seemed to be sleeping already. Felicia fixed her eyes on the landscape out the window, but her lids stood at half-mast. At her feet, a jury-rigged heating element rested in a pail of freshly cut ice.

Every few minutes, the glider let out a soft ping, a gentle warning of an obstacle rising above its preprogrammed baseline of smooth terrain. But this time, there was no ping. This time, the bump rocked the entire glider without warning.

"Whoa!" Felicia yelled out. Supplies fell from an overhead shelf.

"Sorry guys," Hadley called back. She said to Gibson, "What the hell was that?"

"Another one of those mini icebergs, I suppose. But looks like this one reached up to grab us. Damage?"

"I think so. I can't keep this thing on a straight line."

"We've lost a fan. The impact must have taken it out. Great."

"I never saw it coming," Hadley apologized.

"I didn't either, and I was watching. But now we're in trouble. We can't hover any more, not reliably. Next time we hit bottom could be worse."

"We need to use the wheel."

"I don't see any other way."

"The wheel?" Tefia glared through the cockpit door. "I thought we needed to get home fast. Remember, oxygen and all that?"

Gibson grimaced. "We hit something and lost a fan. I'm deploying the rear wheel. We'll skim along the surface, but it's going to be slow going."

"Don't worry, Tefia, I can still go pretty fast this way."

"Good, good." Tefia sounded unconvinced. She headed back to the rear.

Gibson leaned over. "This is a very dangerous turn of events."

Hadley shook her head, squared her shoulders, and urged the glider on. "Don't I know it."

_*_

The team settled down as the adrenaline of crisis ebbed and the cadence of the crippled rover lulled them. It would be a long drive. Felicia seemed to be napping in the back. Tefia busied herself at the fuel cell. Hadley leaned toward Gibson and lowered her voice.

"So we're agreed?"

"On Sterling? I think it's the only way. Confront him with nobody else around and it will be less threatening. And if we're both together, and prepared for anything, he's bound to be more ... " He searched for the right word.

"Safe?"

"I was going to say calm, but yes, safe."

Hadley watched out the windshield, grinding her teeth. "I'm not looking forward to this. It's obvious he's lied about a whole chunk of his life. But I've never been big on confrontation."

"Me neither. But it's for everybody's own good."

"He's not the only one who has lied," Gibson said. She ignored the comment. He added, "What if he's the one who sabotaged our tanks?"

"I don't know."

"I just can't shake this feeling that the key is in the war camps. Sterling was there, or at least associated with them. But I wonder about Amanda. Do we really know where she was in all that mess? I did some digging about the Seven Sisters. The camps. All that dark stuff."

"Bet that was cheery," Hadley said. "I just can't see Amanda as a suspect. Seems like she and Aaron are a package deal, and Aaron doesn't fit. The twin sisters disappeared together, and they were notorious for not leaving each other's side. Aaron throws that equation off."

"Her husband could be an accomplice, a nefarious henchman." Gibson wiggled his fingers, then became serious again. "Think of it: with two murderers it would be so much easier to provide alibis for one another, cover tracks, and so on."

"But they seem so unlikely," Hadley countered. "Neither of them are ripped or menacing or anything."

"Could Sterling be in cahoots with Amanda?"

Hadley shrugged noncommittally. Gibson was sitting uncomfortably in his chair. She could feel the waves of tension wafting from him. "He's much more physically formidable. People do follow charismatic figures or weird causes. Look at the Third Islamic State's Young Jihadists. Or Hitler's Youth Party. And Adolf was a Sunday School teacher compared to Al-Cunelan. Maybe Sterling's involved after all, with his sketchy wartime background."

"Everybody's war time record is sketchy. Not much in the way of intact records."

"But you're right, a robotics expert could go far in wartime," Hadley admitted.

"We'll know about him soon enough."

"And if not him?"

Gibson frowned. "Dakota's too young."

"Is she? Anyone could have an axe to grind, or a blighted past."

"She was just a kid during The War, fresh out of school."

"There are secrets that have nothing to do with The War, you know."

Gibson shook his head. "I just can't see her as the evil villain."

"Villains don't come with curled horns and dripping fangs. They look like the pharmacist or the tax guy."

"Yeah, or the local microbiologist."

"Or a human/robotics interface expert." Hadley's grim countenance matched Gibson's.

Gibson looked at the screen. "How are we doing?"

"Still a long way to go."

II

"Your beepy thing is going off again," Dakota said with mock innocence.

"Sure is," Cann called from the cockpit up front. "I've got the best in side-looking multibeam sonar, so I'd appreciate a little respect."

"That little soprano ping it has doesn't command much respect."

"Proof is in the pudding, kid. Proof is in the pudding. My system sends out a lot more of those rapid pulses per second than just about anybody else's. I can get beautiful detail looking down, and almost as good to the side."

"Why not so good on the side?"

"Different kind of system. But it's still pretty good. I used a similar setup to chart the movements of about twenty humpbacks off the coast of Kauai."

"I thought humpbacks hung out at Notre Dame," she teased.

"Whales, you fool," Cann said dramatically. "Time to get serious; we're less than a klick away from the floor, and that seamount of Hadley's is somewhere up ahead."

"Roger dodger, cap'n."

Cann shut down the front beams. Darkness enveloped them, a deep, aching emptiness to remind them of just how far down in the briny void they were. But as their eyes adjusted, they could make out a faint glow below and to the starboard. The light glowed with an amorphous, cloudlike quality, the red of a burning ember. Cann turned the submersible and descended toward it.

Mesmerized, the two researchers gazed out the portholes in silence. The meters dwindled away, and the light focused into a line, then a glowing snake of smoldering orange along the spine of a ridge. As the little vessel approached, towers of stone rose up to meet it, chimneys growing from the seafloor.

"Smokers!" Dakota called out. "Don't get in those black plumes, but can you take us to the base, around the side?"

"Your wish is my command. I'm just keeping my eye on outside temps."

Cann sailed the craft between two great stone pillars, each capped with billowing smoke. The sonar began to scream, so he shut it down for the moment. At the base of the closest smokestack, cracks jetted more clouds of plume to the sides. Tiny spheres, each the size of Dakota's fingertip, danced and mingled with the billowing fog.

"Are those things moving by themselves?" Cann sounded astonished.

"Can't tell," Dakota said, not taking her eyes off the surreal scene on the other side of the glass. "I'll try grabbing a couple to bring up. I'm putting them in the biohazard containers, just in case we've spotted our first little Europan inhabitants. But I'm skeptical; we haven't seen any active microbes in our other samples yet, let alone something this big."

The globes congregated along one crack, lending a greenish-yellow hue to the rock. Color oozed from the assemblage of spheres, wandering down the sides of the chimney, serpentine lines of rust and green against the gray stone. "Looks like bleu cheese," Dakota said, taking another sample.

"Stop. You're making me hungry. Shall I take her around the other side?"

The craft banked slowly around the column of multi-colored stone. As the nose turned back toward the tower, the sub's light beams crossed a rich colony of the spheres. Here, something more complex was taking place. The globes at the edge of the crack seemed to be waving tendrils into the plume, while the ones adjacent held the grouping firmly on the side of the volcanic pile.

"Now, that looks like symbiosis," Dakota said, barely hiding her excitement. "We've got good imagery, good samples."

"We've also got some pretty good heat," Cann added. "My hydraulics on the volcano side are redlining. I need to pull back."

"Sure, sure," Dakota said, distracted by the abyssal wonderland outside.

An hour later, the team of two had sampled three sites containing a variety of spikes, spheres and star-like spots draped across the geyser pedestals.

"Just looking at them, nothing seems to be obviously swimming," Dakota mused as Cann began his ascent. "But geological forms don't usually do the things we saw down there. This is cool."

"Just what you came for, yes?"

"Still waiting for those whales," she said.

"I'll fire up the side-looking sonar again, and the other for our navigation to topside."

The craft ascended fairly rapidly, but in the darkness of the open water, Dakota had a hard time sensing movement or speed. It was quiet, deathly still save for the environment fans. Suddenly, the sonar sprang to life.

"Something big out there," Cann said.

"Out where?"

"Port side. Can't get a fix on it."

A recorded voice stated flatly,

COLLISION   ALERT … COLLISION   ALERT … COLLISION ALERT …

"Oh no you don't," Cann told his control panel, bringing the craft to a near standstill. The sonar continued to ping, faster now.

"Sounds like something is coming," Dakota said.

"Wishful thinking, my dear. It could be a ghost, a sort of acoustic mirage. Happens all the time."

But both of them were looking out the widows. Dakota strained to see. The darkness pulled on her eyes, a physical presence as relentless as the vacuum of space. Out there in the distance, she saw a flicker of light.

"Did you see that, Jeff? That light?"

"Might have just been a reflection from our own lights."

"Yes, or bioluminescence."

Another flicker of color blazed in and out, followed by several more. Were they in a line, or was she imagining it? Then she saw it: a dark, long shape undulating through the blackness, even darker than its surroundings. She struggled to make out a shape, a fin, a tentacle, anything.

"Did you see that? That long thing?"

Cann paused.

"Well, did you?"

"You know, our brains often go astray as they try to figure out confusing input. We might perceive separate light sources as chained together, like aircraft in a night sky, for example. Our minds are programmed to see patterns, even when they aren't there. They fill things in where there's nothing but empty space."

"But the sonar!"

"Yes, I know. Very intriguing. But sonar can be fooled, too. We get scattering from stuff floating around in suspension."

"Like a car's headlights in the fog?" Dakota tried to hide her frustration.

"Exactly. And the way a target reflects sound, that's called target strength, can be thrown off by echoes from the background or from junk floating between us and whatever it is. Or a gradient in water salinity, shifts in temperature, the odd surface bounce from a skewed ice shelf. We'll figure it out.

We've got plenty of data to study topside, but not plenty of O_2. Time to head home."

–*–

Dakota let herself think about the things she shouldn't. Sweat broke out on her forehead, puddled in the small of her back. She lifted her chin, trying to let the air blow down her neck, across her chest, through her hair. She thought about the darkness. She thought about the tons of water pressing in on every inch of skin on their little red torpedo. She thought, most of all, about the depth, the distance up to the open air and light. It was so cold and dark out there. So very dark.

III

"Disappointed?" Jeff Cann asked, helping Dakota from the rear of the submersible. Sterling stood by at the winch.

Dakota took in a deep breath, feeling the humidity from the moisture dripping off the sub, a spring shower filling the room with its aroma. "Not at all. It's beautiful down there. Besides, with all those organics floating around, there must be something bigger. We just have to find it. And those big pings on your sonar just might have been something swimming. Something big, or a bunch of little somethings."

"Such a turn of phrase. You should be a playwright."

"Yeah, well, the fact is, I've seen schools of minnows with a similar sonar signature."

"With luck, you and Amanda will both get another crack at it before we have to pack up."

Amanda burst into the room. "Did you see it? The sonar?"

"Let's everybody relax," Cann held out his hands, moving them as if pushing down a balloon. "Sonar is a tricky thing, and with all those irregular ice surfaces, it could have been anything."

"My vote is a giant squid," Sterling said, stowing some dripping equipment.

"Oh, yes," Amanda said. "I'd take that. Arrange it for us, Sterling."

"Before our next dive, what I *can* arrange for is a short-range sonar on another one of our ROVs. I had one mounted before, but it was a brief reconnaissance, so maybe we just weren't deep enough. We can send ROV 2 down; might dig up something."

Dakota grinned. "Let's. Maybe we can bag our Europan whale before the others are back."

"We'll have to make it soon," Sterling said, looking at his watch. "Data from the glider shows that the gang is on their way back from their precious geysers as we speak."

"I wonder why they haven't called," Cann said.

Amanda frowned. "Should we be worried?"

Sterling shook his head. "I've been monitoring their suit location beacons. I was even able to get sporadic med readouts this morning, using the Europasat. They've all had plenty of activity. They just aren't saying anything. Maybe it's some kind of hardware failure. We'll know soon enough."

13

Return

I

Memory could be a funny thing, Dakota reflected. A whiff of something, a perfect tone, a simple phrase could bring back something from the past. It was a phrase that did it for her, a phrase she hadn't thought about since the trip out.

"Together for strength; together for power."

Amanda had said it, or was it Aaron? But it was a quote from the Eastern Alliance. One thing Dakota Barnes knew was her history. She loved reading the glyphs of the Mayas, thrilled to the narratives of the terrifying hordes of Genghis Khan, hummed along to the ancient songs of the Renaissance or the gay nineties. And this phrase had stuck. Not ancient history, but recent events. In her short life, a decade was a long time, but in the grand march of humankind's business, it was a breath, the blink of an eye.

The phrase carried so much emotional and historical baggage. The prime minister of the North American Union, Jorge Devereaux, had even declared the phrase to be a "slur against civilization." Why would anyone use it in any social setting? But the remark had been offhand, reflexive.

What were Amanda and Aaron up to?

She waited until everyone had assembled for dinner. This was the perfect night, with fun and games on the schedule. She made an appearance, gave it a few minutes, picked at her food, and then made her announcement.

"Well, I'm going to take me and my headache to bed."

"Would you like some pain meds?" Aaron offered. "I've got these great little blue pills."

© Springer International Publishing Switzerland 2017
M. Carroll, *Europa's Lost Expedition*, Science and Fiction,
DOI 10.1007/978-3-319-43159-8_13

"No, thanks anyway. A little sleep will work wonders."

"But it's card night," Jeff Cann whined, pulling a deck from his front pocket. "I was going to do some serious payback on you for that last round of Hyperhearts."

"Some other time, loser," she mocked.

"Would you like two of us to escort you home?" Sterling asked. Everyone laughed.

"Yeah, like that's going to work," Cann said. "What kind of policy is that?"

Dakota shrugged. "Maybe Hadley will explain her great wisdom to us when they're back."

Everyone bade her goodnight. She didn't have much time. She made her way quickly to Aaron and Amanda's quarters. Its proximity to the diners gave her pause, but there would never be an opportunity like this one.

The room was dark as she entered, and she kept it that way. She flicked on a small flashlight. She had no idea what she was looking for. She could feel a bead of sweat snake its way down her back as she felt through the blankets on the two bunks. Her eyes scanned the room, desperately looking for anything out of the ordinary. But it was a frugal living area. If they had any personal trinkets like photos or hangings or even personal monitors, it was all back at Taliesin. That only made sense. They had all been encouraged to bring the least amount of personal belongings possible on the treacherous expedition to Europa's southern arctic.

She turned toward the door to make a quick exit. Raucous laughter reverberated down the hall. Her foot brushed against something under a wall conduit. She leaned over, reached under, and pulled out a box. It was about half the size of an attaché unit. A single silver clasp at the front held the hinged top closed. She twisted the clasp and swung the top up. Inside, blue foam filled the case. That foam held an indentation that chilled her blood. The hollow outlined the perfect profile of a handgun. Below it, inset into the foam, nested a row of five empty depressions, each an identical outline. At the end of the row, two more of the depressions still held their treasures: clear vials filled with some type of liquid. A covered needle extended from the top of each.

"No you don't!" a voice resounded down the corridor. It sounded like Sterling. "You can't abandon us when you're ahead like that."

"I won't be gone long. Patience, my boy." That was clearly Amanda's.

Dakota slammed the case. She shoved it into place with her foot.

"She's got to know," she mumbled. Desperately, she scribbled a note on a scrap of paper, stuffed it into her belt, and ducked out of the room. In the darkened hallway, she heard footsteps. It was too late. She flattened herself against the wall, but it was no use. Amanda would pass right by her. Changing tack, she began to walk toward the card game. Amanda met her a few paces from the room. Dakota put her hand to her forehead.

"Hey, Amanda, I think I may take Aaron up on one of his little blue pills."

Amanda glanced in the direction of her room, and then let her pass toward the gamers. "Yes, you get something for the pain, poor dear. Sweet dreams."

II

"How far?" Gibson asked urgently.

"Getting near," Hadley said, looking at the rangefinder. "We've got less than twenty clicks, maybe half an hour. Still no radio?"

Gibson shook his head. He craned his neck and looked back toward the fuel cell. The bucket overflowed with ice chunks, but his crippled contraption could only go so fast, and it wasn't keeping up.

"O_2 is pegged at zero," Hadley called out. "Everybody put on your helmets."

The last 10 kilometers seemed an eternity, but the dome finally reared up beyond the icy hummocks ahead.

"This was close," Felicia said.

"Gibs, try the radio again."

"I just did. Nothin' doing."

"If I get a bit closer, they can pick up our suit radios. Tefia, got your helmet ops running?

"All secure."

"Try the common channel and see if they can shut down the field to let us in."

"Will do."

"Okay gang, we might as well seal our suits and head for the airlock as soon as they let us in."

"South base, this is Glider One. Do you read?"

"Roger, Glider One. We've been expecting you, for some time!" came Jeff Cann's voice over the static in their headsets. The marker lights of the magnetosphere generator began to fade off. "You didn't call, you didn't write, you didn't send a postcard…"

"We've had some trouble with, ah, communications, among other things."

"Yeah, well, we're having some trouble here, as well. Come on in and you can help us."

Hadley looked at Gibson. She took the glider in, skirting a pile of crates and settling next to the airlock. Everybody piled out and scampered to the lock. As the air flooded the chamber, so did relief.

"Room to breathe," Felicia said.

"Oxygen to breathe," Tefia added.

"Three cheers for fuel cells," Tefia crowed to Gibson.

"Aw, shucks."

Hadley said nothing. Gibson's general attitude still stung. She was conflicted: should she have it out with Gibson in a major argument, or let things simmer as they were? The stakes were even higher in light of the glider sabotage. But she needed to defend herself, to make him see that she would never, ever put the lives of her team before science results.

A line from Hamlet tickled the back of her mind: *The lady doth protest too much, methinks.*

The hatch slid open. Jeff Cann and Sterling Ewing-Rhys stood side by side, looking worried.

Hadley stepped forward. "Gee, I knew you guys would miss us, but this is ridiculous."

"Yes, well," Sterling's eyes darted from side to side, "Welcome back."

"Out with it, guys," Gibson prodded. "What's going on?"

"We can't find Dakota. Or Aaron."

Hadley slid her hands to her hips impatiently. "Where's Amanda?"

"On comms, trying to raise either of them on the common channel."

"Where have you looked?" she barked.

"Pretty much everywhere."

She glanced toward the submersible area, with its bottomless pit, and then thought better of it. "Then they must be outside."

"What would they be doing—" Sterling's voice trailed off as he gazed out the window.

Hadley spotted it, too. Just beyond the small ridge to the east, the second glider sat where it had been parked since its arrival. Across the side window, a spider web of cracks spread across the plex. A hole at center was venting a snowy fog into the surrounding vacuum.

"Jeff, grab a patch kit and meet us out there. Gibson and Tefia, you're with me. Sterling, get a suit and come out with Jeff as quickly as you can."

Gibson, Tefia and Hadley lunged back into the lock, sealed their suits, and flushed the air from the chamber. Hadley forced the door before it was completely safed, and dashed to the glider. Leaning her helmet against the glass of the windshield, she peered in, shielding the sides of her head with her hands.

"They're in there, both of them. Let's go!"

As they entered the lock, Sterling and Jeff Cann arrived. They all piled into the cramped lock together.

Gibson checked a readout on the interior hatch. "There's still pressure in there ... some."

"What were they doing out here?" Tefia mumbled. Through the frosted glass of the airlock port, they could make out the two figures. Dakota lay across the floor, her helmet on the seat beside her. Aaron draped over the back of a chair, his arms hanging limp. His helmet was still on, but the visor was open.

Jeff Cann popped the hatch violently and dove across the cabin to Dakota's side.

Tefia took off her gloves, pulled out a monitor and checked Aaron. "He's dead."

Hadley's voice held the tone of command. "Everybody keep your suits pumped up. Let's shut off the O$_2$ until we can seal that window. We can't afford to waste the air."

Jeff Cann called out, "Wait! Dakota's got a pulse."

Hadley asked Gibson, "What's our pressure?"

"Five point five psi and falling fast."

She reached along the base of a seat for a metal locker door, pried it off its hinges, and jammed it against the window. "Somebody turn up the air until we can stabilize Dakota."

Gibson forced a small sheet of plastic over the metal to help seal it, but it wouldn't do for long. The air screamed through the lousy seal. The window had a slight curve to it. A flat plate of metal would never do. Hadley surveyed the interior. What could she use? Her eye settled upon Aaron.

"You're sure he's dead?" she hollered at Tefia. The doctor nodded. Hadley pulled his helmet off, yanked the locker door away, and jammed the rounded form into the hole. The leak quieted, but the glider was still losing pressure.

Tefia reached inside Dakota's collar ring and gave her a shot of something. She grabbed Dakota's helmet and slid it back in place, then turned her suit oxygen up.

"We need to get her inside, stat," Tefia said.

Hadley pointed to the lock. "Gibson, Sterling, you guys help Tefia wrestle Dakota back to the medlab. The rest of us will take the next lock cycle."

With the others safely through the lock, she glanced back at Aaron's body. "Suits!" she called. She sealed her helmet, pulled the helmet from the window and dialed the air pressure down, letting it make its way back to zero.

With Sterling otherwise engaged, Hadley could relax and survey the compartment. Both of them had been in here long enough for Dakota to remove her helmet. Some kind of struggle ensued. But what kind?

While Jeff worked on sealing the window, she scanned the seats. Shards of safety glass lay on the seat below the breach, pieces of the inner layer of glazing. It would take a great deal of force to break through it like this. She turned her attention further toward the rear. There, on the floor next to the seat where Aaron sprawled, lay a small black object. She stepped around Aaron's corpse and approached it. It looked like a toy gun. It didn't seem to be made of the right stuff for a pulse weapon or a conventional firearm. This gun was designed for something else. On the floorboard next to it, a spent vial had rolled against the bulkhead. She had seen a vial just like it back when she was dating that wildlife management grad student. There was no doubt in her mind: this was a dart gun.

_*_

Sterling and Gibson set Aaron on a gurney in the rover docking bay. It was the closest thing to a morgue that the outpost had. They closed the door for privacy; Amanda deserved a few moments with her late husband. Hadley tried to hear something behind the door: soft sobbing, sniffling. But Amanda kept silent vigil. The two men stood outside the door, aimless pallbearers in the aftermath of a funeral. Hadley turned and walked to the little medlab.

Tefia lifted Dakota's eyelid and shone a flashlight into it. Then she waved a handheld scanner across her forehead. Jeff Cann sat at Dakota's side, holding her hand as if it was made of spun glass.

"How is she?" Hadley asked.

"Critical, obviously. She's got those same blisters."

"They're from this," Hadley said, holding up the empty vial. She looked to Jeff for some reaction, but he was engrossed in gazing at the patient.

"What is that thing?"

"Delivery system for a dart gun."

Tefia nodded. "Now *that* makes sense."

"Can you tell what was in it?"

"Not here. We don't have the equipment. If she's to stand a chance, we've got to get her back to Taliesin."

Hadley began to pace, thinking aloud. "Jeff patched the window, so we can all pile into Glider Two. The primary is shot. It would take us a day to fix its oxygen tanks, and I'm not even sure that vertical fan can be resurrected."

Tefia shook her head. "No good. It will take too long to drive back. I'm calling for a jumper."

"Would they have room?"

"Taliesin has one twin-rider version. Hopefully it's not under repair or some such nonsense."

"Don't people have to stand up in those?"

"We'd have to prop Dakota up, maybe bind her to a stretcher and put her in that way."

Amanda thrust her head through the hatch, looking more distracted than sorrowful. "What's our patient's status?"

Tefia shook her head and said, "Not good."

Amanda glared in Dakota's general direction, and abruptly left without waiting for details. Hadley leaned in to Tefia. "We need to have someone watching her at all times."

"I will be."

"No, Tefia, I mean *constantly*, in shifts. Either Gibson or me or Jeff Cann."

"Seems a bit melodramatic."

"Consider it a personal favor."

"Sure, you're the boss," Tefia said, glancing at the comatose biologist once again. "We've got to get help."

III

Tefia leaned in toward the comms unit as if a few less inches might make things clearer. Hadley gazed over her shoulder, hunched in the same direction. "I *said*, do we have any Epidromapril?"

On the screen, Taliesin's only med tech on site, Ben Reynolds, shot back, "Yeah, right. I keep it in the dungeon with our other medieval tools."

"Sometimes it's used in conjunction with Topalazine. Check the labels. If so, send me about a dozen doses."

"Why in the worlds would you want Epidromapril?"

"It's an old war trick."

"Before my time."

"Just trust me. We need this stuff stat."

"Epi-friggin-dromapril," Benjamin mumbled as he signed off.

Hadley frowned. "Will it be soon enough?"

"I don't know. They're sending a jumper on its way before the hour's out."

"What's this old war trick?"

Tefia paused. "I had a hunch early on, but it was so crazy I didn't want to say anything until I had some proof. Sadly, now I have some proof. These symptoms that our dearly departed share with Dakota are similar to the results of a toxin used by the Eastern Alliance during The War."

"And this weird drug you asked for?"

"The only known antitoxin. By itself, it's been out of use for a long time. But they still combine it with some compounds. That's what I'm hoping for. But this late, I don't know if anything will work. There's something else I should tell you."

Hadley stared at her, waiting.

"It's about Aaron. He died quickly, of a subarachnoid hemorrhage."

Hadley blinked.

"Head trauma. Basically, the brain tears away from the skull and you get bleeding into the membrane around it. Not before he got off the shot, of course, but I think Dakota put up a really good fight before he was able to shoot her. Looks like a kick-injury to me."

"A kick to the head? Pretty hard to do in a pressure suit."

"I'm pretty sure she knocked him down first. A fight like that wouldn't explain the window, though."

"This would," Hadley said, holding up a small caliber bullet casing.

"Old school. Find a gun?"

"I'm sure we will." Hadley looked across the habitat at Jeff and Dakota. She turned to Tefia. "Fingers crossed for those drugs. There's something I've got to do." She stormed out, a woman on a mission.

IV

Hadley entered the bay with an uncertain Gibson in tow. At the center of the chamber, the cherry submersible hung on its cable above the shaft to the abyss, frigid steam billowing up around it. Sterling studied a screen, monitoring an ROV that was pinging sonar somewhere down there. He wore a headset for direct interface with the little submersible, its electrodes impaling his skull. He was alone.

Gibson gawked at Sterling, and then at Hadley, a look of understanding crossing his face. "You could have told me where we were going," he hissed.

"I was in a hurry," she said aloud.

Sterling pivoted. "Oh, hello. I'm … I'm working. It's how I deal with stress, I guess. I'm sorry." He sat heavily on a stool, looking at the floor. Sterling looked wrung-out, beaten down and exhausted.

"We can make this really efficient," Hadley said. "You can be honest out of the gate with us, or we can play lots of gee-what-are-you-talking-about games and breathe lots of precious oxygen while Gibson and I wait for you to admit it all. You know which approach I prefer."

Sterling sat up straight. "Speaking of games, what is this, *Truth or Dare?*"

"Look," Gibson said. "A lot's been going on around here, and we need to know all the facts. We know you lied about your history over the last few years."

"Your time on Mars," Hadley added. "It doesn't add up, and under the circumstances, that is a very bad thing. That makes me very unhappy."

"And you don't want your expedition leader unhappy," Gibson warned, an edge to his voice.

Sterling wilted against the counter.

"I didn't want it to come out this way. I thought I could be a good scientist and prove myself and then let you know after."

"Let us know what?"

"That I was institutionalized. For a time."

"Institutionalized?" both Gibson and Hadley said in unison.

Sterling looked at them as if they were small children who needed an explanation of crossing the street. "As in the looney bin. Mentally...unstable. That's me. Well, that was me. I definitely needed to be in there after The War. I checked myself in when the nightmares wouldn't go away."

Suddenly, Hadley's entire perspective shifted. Perhaps Sterling wasn't predator, but rather victim. "Sterling, that's not looney. That's PTSD or something."

"Yes, there are lots of diagnoses, and I've tried out quite a few. You're right, of course. But as the patient, it's my prerogative to be politically incorrect about my experience."

"I'm sorry," Gibson said. "It's nothing to be ashamed of, certainly. An entire generation suffers from it to one degree or another."

"Another war to end all wars," Sterling observed. "My voluntary incarceration was a growth experience. And not all bad, I can tell you. I am a different person now, and better for it. But its impact on my career was difficult. I was out of the mainstream of research for nearly four years, with the exception of a really poorly constructed paper I wrote early on. I attempted to keep up with things from in there. Subscribed to all the important journals. Even had a visit or two from colleagues. But they came with the understanding that mum was the word, that as soon as I got out I'd be back in the saddle, doing my research."

"How did that go?" Hadley asked. "Once you got out?"

He smiled. "Fooled *you*, didn't I?" His expression fell. "I am sorry, Hadley. I know this is grounds for defunding."

"Look, as far as I'm concerned, you do the work, you get your full grant."

He looked as if he was going to cry, or as if he was going to hug her. She sent a look his way. It wasn't anger, exactly, nor derision—but it carried a warning. "Then that's settled. But there's something else that maybe you can clear up for us." Her voice was steady, but her heart thudded in her chest. She grabbed a chair and sat down. Gibson did the same.

"You see, Sterling, we know that's not the only misinformation you supplied," Gibson said.

"What are you talking about?" Sterling asked, his patience apparently waning.

"The war years," Hadley said.

"I thought we had gone beyond all that," Sterling objected.

"Not me," Hadley shot back. "This gets a little personal. My father was in one of the Antarctic camps."

Sterling peered back at her with a hint of indecision, of caution. He spoke slowly, in measured cadence. "I see. I am sorry he was. Did he…" He hesitated.

"Survive?" she filled in the blank for him. "No, he did not. And you were listed in the manifest as part of the transport system."

"Not of people," he said defensively. "Never people. Only supplies. I had no idea. Lots of people had no idea what those camps were until after. Those nightmares I told you about? Some of those came from seeing the records. The bodies. The slaughter. My God." He fell silent, looking off into some past horror. "Lots of us didn't know. Lots."

Hadley leaned forward. "It's true, Sterling. I didn't know what my Father must have gone through until I saw those images. I had nightmares, too. We all did."

She stood, turned her back on him, and left.

14

The Odd Couple

I

Hadley cradled her head in her hands, elbows on her knees. The habitat quarters seemed stuffy, claustrophobic. "I'm just numb."

"Me, too," Gibson said, keeping his distance. "And a bit panicked. I'm telling you, Had, we obviously need to tell Tefia what's going on. The full deal. Maybe she can watch our backs."

"But Dakota got the best of her attacker."

Gibson frowned. "Why are you so insistent on this? I just don't get it." He was fuming. He quieted for a moment, and then said, "There were incidents where Aaron had an alibi. Which might mean it's between Amanda and Sterling."

"An extremely peculiar couple."

"I don't think so."

Hadley looked up. "You don't think they make an odd couple?"

"I don't think it's down to the two of them. Not Amanda and Sterling. The more I think about it, the more convinced I am that it's Sterling, and somehow he was involved with Aaron in the whole thing. He goes all doughy-eyed and whines about being the patient, locked up in a ward, but I just don't see it."

"Easy enough to check."

"That's not what I mean. Sure, he was in a facility. I'm sorry he had to go through that, but I'm skeptical. Has he really recovered? Is he back in the main population because he was cured, or because some bean-counter dismissed him after his insured amount of treatment?"

© Springer International Publishing Switzerland 2017
M. Carroll, *Europa's Lost Expedition*, Science and Fiction,
DOI 10.1007/978-3-319-43159-8_14

"Maybe there's more cluttering his attic than cobwebs," Hadley suggested. "Maybe he fooled everyone. Maybe Sterling's been a ticking time bomb all along."

"It's conceivable," Gibson conceded. "Rather than some conspiracy or war revenge, we may have a serial killer on our hands. Just pure, simple, sociopathic insanity. Still..."

Hadley kept exploring the point. "And he's a big guy. With the element of surprise, he could have easily overpowered Orri or Joel. From a purely physical standpoint, I guess I'd take him over dainty Amanda any day." She studied Gibson. "I know that look. You're not so sure, are you?"

"Two things. We know Aaron didn't work alone, so that precludes the serial killer scenario. And something else has been bothering me."

"What now?"

"Dakota's attack just wasn't a consistent MO."

"How so?"

Gibson put his hands together, organizing his thoughts. "All the others died in isolation, but not in remote locations. They were all found in fairly familiar surroundings—air locks, bedrooms, labs. Places where their guard was down—places close enough for the killer to make a quick exit. But not this time. This attack smacks of desperation, of changed plans. Why would Aaron lure her out to the glider?"

"Assuming she didn't just stumble upon him, which I doubt. We're not in a sprawling civilization any more. We're in a secluded, contained location. Maybe he needed to get her away from the others in order to not get caught." She rubbed her eyes with the heels of her hands. "The whole thing is a tangle of events and opportunities and...I just wish I could see it. I wish I could visualize all the pathways, trace them back to something."

"You're a good scientist. You don't need to kick yourself for not being a great detective, too." Gibson put a hand on her shoulder.

"Gibson, I'm sorry. About everything. About the way I—"

"You can be awfully stubborn. I just think we should have, you know..."

He was still not over it. Still. He could be impossible. "Look, we've got to figure out a few things concerning more current events, like who needs to be locked up before they kill again."

"Before they kill *us*."

Hadley stood and stiffened. "That's it. We've got to tell Tefia."

"Finally. Sanity reigns."

Hadley held up her hand. "Just her. Everyone else needs to be concentrating on the work left. We've got a short window of time before we go back. But if we have Tefia helping us watch, maybe they won't have opportunity. We keep

Amanda and Sterling separated, and watch each other's backs. We simply need to deprive the murderer of the chance to knock off anybody else."

Gibson's face darkened. "That's been working well so far. Sounds so simple."

She pursed her lips. Her face colored.

Gibson said, "We've got to locate them first. Where is everybody?"

"It's a small place. We'll find them. Let's go. We've got things to do."

II

Sterling Ewing-Rhys knew he was smart, despite the years of taunting. The spiteful bullies on the playground, the managers who passed him up unfairly, even the psychiatrists at the facility on Mars. None of them understood how clever he really was. They had come to appreciate him during The War, of course, but the way things had gone, he had lost any bragging rights. Those were for the victors' side. But now he had the confidence that he was right. He had proven it, if only to himself.

Yes, he had worked out a great mystery, something nobody else had. And now he would cash in. Not in a monetary sense; he wasn't interested in money. He simply wanted to show them, show her, that he knew. He wanted to hear her say, "Well done, Sterling. How did you know?"

She was due at the dome in a few minutes, and he wanted to be waiting. He made sure he was early. He walked into the darkened storage area. It seemed strangely warm. He loosened his collar. As he did, he sensed that he wasn't alone. Just as he began to turn, he felt the bee sting on the side of his neck. He went down fast, and as he fell, he knew what had happened. He ended up on the floor, sitting with his back against the wall. To his surprise, he wasn't dead yet. That would come soon enough. But he could actually think, and even move his mouth.

He could move his eyes, too, just enough to look up at his assailant. She held a little glass tube horizontally, the aftermath of stabbing Sterling's neck with it, dagger-style. She dropped it onto the floor, where it shattered, leaving a needle lying on the ground. He struggled to move his lips. "I knew ... I knew it was you." His tongue felt like a wad of cotton. "The years ... have been good to you, ironically."

"I've had some, ah, additions, shall we say? On a genetic level. Softens the features, changes the cheekbones a bit." She began to walk around him, circling like a vulture awaiting a meal. "We were surprised to see you on the ship. Gratifying that you didn't recognize us right off."

Sterling tried to nod, but his head wouldn't move. "Yes, yes," he slurred. "But it was your sister. I recognized her. Even with facial hair. And the voice. Lower, yes. But still."

"She was the one who made the real sacrifice. The hormones. The surgeries. But she wanted it that way, and she always got what she wanted. Even after The War."

He felt hurt, betrayed. He could tell his body was trembling now, but he couldn't feel anything. His vision simply wavered with the shaking. "Why did you do this? To m-me."

"We decided we had too many loose ends these days. At first, I figured you had as much to lose as we did if someone found out about your past war record. But Hadley Nobile is digging around, putting some facts together. And then Aaron blew it with Dakota, and that forced my hand."

"I just wanted … to join you … to … " He coughed.

"No matter. Too late now, my dear. It was good with you, in the old days. But life is more complicated now. You'll be happier."

Sterling could feel himself sliding sideways, his head moving toward the floor. The ghostly figure patted him on the shoulder and began to drag him toward a storage bin. She would be the last person Sterling Ewing-Rhys saw alive.

III

Sterling's ROV 2 had done well. Its data safely recorded on the mainframe, Jeff Cann prepared to call it back in. He sat at the monitor, nudging a joystick. The little robot indicated that it was turning and beginning its ascent. The return would take an hour up to the cable, and then another little while to reel it in the rest of the way.

Just as Cann stood to leave, he heard the sonar ping. He had meant to turn it off to conserve power, but now it was beeping away with ever-increasing frequency. Something was out there. And it wasn't a school of minnows. It had a fluid movement, coordinated, graceful. And it wasn't just drifting around out there. It was approaching.

Cann studied the monitor. The automatic imager continued to take frames, once each few seconds, as a matter of course. The simple ROV had no other instruments mounted to it for this reconnaissance. He should get Sterling, but he couldn't move from the screen. It had him in a paralytic grasp.

The pings became louder and increased in speed. Now, they sounded like the patter of rain. Low in the field of view, something moved. It was a single object, and judging by the sonar, it was immense, the size of a transport. More vast than a blue whale. Seconds later, the next image brought it more into focus. It was moss-green, mottled, bulbous. The next frame revealed more detail. Were those scales? Rows of knobs? Whatever it was, the leviathan was turning. The next view showed the rounded head from the side. Trailing behind it was a long wall of greenish body, with rows of dark and light spots, tiger stripes, and glowing bioluminescent openings in a long lateral line on its flank. The little lines of identical shapes were something that a marine biologist, someone familiar with blotchy whales and spotted anglerfish, a biologist like Dakota Barnes, would have loved.

Frame after frame now came in. The wall of swimming flesh continued to scroll by the camera's lens. On and on it went, undulating and arching like a great eel. Cann could see the outlines of rippling muscle tensing beneath the flesh. What organs pulsed within those tendons and sinews? What fluids coursed through those alien veins? This was the Everest of the expedition, the summit of hopes for their combined discoveries, and no one was there to share it with him. Cann reached for the intercom key.

The body continued sailing by, a train of life in the abyss. It began to taper. Suddenly, a great fan-like fin appeared. As the creature turned away, the fin twitched, sending a tsunami of current toward the submersible. The ROV banked and wobbled, trying to right itself.

Abruptly, the probe fell silent. Cann commanded it to respond on the emergency backup channel.

Nothing.

The data feed had died. ROV 2 was no more.

Cann keyed his wristcomm. "Sterling, get down to the bay right away! Sterling?"

But Sterling's feed seemed to be just as dead.

15

Blasts from Those Pasts

I

It was an article. A very old news item, something that should have fallen away from the servers decades ago. But there it sat, a forgotten bit of digital flotsam from the post-war era. News from his hometown. From the town where both he and Hadley had come of age before the war years.

Local Engineer May Have Died as Victim of Seven Sisters

As far as Gibson could tell, the piece was not well researched. It posited and proposed without citing much firm proof. But what it suggested reinforced Gibson's fears: that the death of Hadley's father somehow linked up to the Seven Sisters. No wonder she hadn't volunteered the fact of her father's war involvement to the others. Hadley had personal reasons for going after the Seven Sisters, and one of them just might have been in their midst. Hadley Nobile was truly a woman of mystery, and he didn't like it. What part did her obsession play in this expedition? What risks had she taken, and with what motives? He felt like screaming at the wall. At her.

He remembered some old quote about the "sins of the fathers" being passed down, generation to generation. Umberto Nobile and his expedition certainly haunted Hadley. And her parents' war experiences would give anyone nightmares. But family history seemed so distant, so removed from the events of the present. Was there something else, something more urgent, urging her on? He suspected there was, and he knew how to find out.

© Springer International Publishing Switzerland 2017

M. Carroll, *Europa's Lost Expedition*, Science and Fiction,
DOI 10.1007/978-3-319-43159-8_15

Back into his records he dove, back into the war years. Orri. Ted. Dakota. What were they doing? Where had they been? At first blush, Orri's bio seemed to jibe, but Gibson took a tangential approach to some historic footnotes and found what he had been looking for. Orri Sigurdson had been a guard at a mining camp on Pallas, one of the camps that supplied the Eastern Alliance. Was there a tie-in with Hadley's father?

He confirmed his other hunch even more quickly, but the results baffled him. Ted had been at Perth, Australia, the great processing center for prisoners en route to the Antarctic camps. Gibson suspected Ted's involvement might have been more than spiritual. And the common factor, of course, was Hadley, the one who had hand-picked the team. Every one of them. What could it mean?

His monitor chimed. He tapped to accept. Amanda's face snowed into focus. "Gibson, I need a good geologist for some advice. I'm wondering about ice flows and their interaction with our subsurface sea. I'm thinking we might be able to do some creative studies together. Can you come to the storage bay?"

He didn't know what to do with his revelation. Perhaps a change would give his mind a chance to process, to strategize. His suspicions aroused, he said, "Do you know where Sterling is?"

"He's in here with me." He relaxed. Safety in numbers. Maybe Hadley had been right about trios.

"Sure. Why the bay?" But she was gone. Maybe she didn't want to disturb the others. They would all be in the dome or in the lab, preparing equipment to install on the last flotilla of ROVs. The submersibles were scheduled to deploy one last time, at midnight. That kept everybody pretty busy. Best to leave them to their devices. Perhaps Amanda had come up with a variation on the idea of finding flash-frozen remnants of life on Europa's surface, at the sites of upwelling or venting. To date, none had been successful, but she was clever.

He stepped into the corridor, glancing both ways. Cold sweat broke out on his forehead. He looked at his wrist comm. He tapped it. "Hadley, you there?"

Silence. Maybe she was in the bathroom. Why didn't she answer? Should he go back, tell her where he was headed? But if Sterling was there, he had nothing to worry about from either of them. The pair offered a safe situation by simply being in the same room. He made his way toward the lonely storage bay.

II

The words burned the surface of her monitor. While the others prepared for the final ROV deployment, Hadley took a moment to dig into the past again, this time with added enthusiasm, and a specific goal. She had shut off her wristcomm and personal networks. She needed to concentrate, and she didn't need to be nagged by Gibson. His tunnel vision was getting old.

It didn't take long to find the phrase she was looking for, a phrase that had not been spoken aloud in a very long time, words that haunted a generation.

"Together for strength. Together for power." Words scrawled on a piece of paper under her pillow. Dakota's handwriting. Hadley remembered armies of followers of the Eastern Alliance, automatons, mouthing the words at great public assemblies. And now, she remembered Amanda saying them at dinner all those weeks ago, on the way out from the inner system. "Together for strength," she had said. And Aaron had responded reflexively, without thought. At the time, she hadn't quite caught what Aaron had said, and the conversation veered off in other directions, but now she thought she knew. "Together for power." What had happened after that? She had not been paying attention. How she wished she could conjure up that scene again. Did he look panicked? Embarrassed? Did Amanda scold him with a frown?

But the second article on her screen alarmed her even more. It detailed the personal information—what little was left—of the Seven Sisters. The twins were the ones who were still missing. Authorities conjectured that they had escaped, possibly to Mars. And the one who had been a geneticist went by the name of Erin.

Erin. A twin sister.

Aaron. Was it coincidence? She was beginning to think there were no coincidences. Something itched at the back of her mind, something Ted had said. "I could hear it. In their voices. They're the same. Listen, Hadley. Hear."

She knew there were few high-resolution images of the Seven Sisters, except for several INTERPLANPOL snapshots caught in later years. Fewer still were the audio recordings. Audio degraded more quickly over time than the visual stuff. She pulled up a recording of Amanda from her expedition dossier. Then, she tabbed over to a series of files of Aaron's voice. She logged onto the skein, historical section. She had remembered one speech by one of the Seven Sisters at a rally, a famous speech. She fed the file into a simple sound-mixing program. Its distinctive lines painted mountains and valleys across the screen. She paused the file, and then overlayed the one from Aaron. The lines were no match. She lay Amanda's across the screen, and shifted them laterally. She dragged them, one to the other, until they overlapped. Like the overlays of an anaglyph, the images duplicated to make a greater whole. "The voices," Ted had said. "They're the same." And they were.

The song of the meadowlark. Hadley had to get to Gibson.

III

Gibson could hear the voices of busy scientists as he passed the laboratory door. They didn't need a geologist like him, or even a volcanologist like Hadley. Not now. Now was the time for people who knew ocean currents and remote operating vehicles and submarine robotics and, hopefully, biology. But they didn't even need Amanda until the plucky little robots made it down into the open water.

Biologists. With an abrupt ache in his chest, he wondered how Dakota was faring. The jumper wouldn't be here for another two hours.

He continued down the corridor, anxious to know what Amanda's proposition was. As agreed, Hadley would follow soon, but she had to finish some kind of program run. She said she would be "right behind." And there would be people next door, in the dome area. Both Sterling and Amanda there. As his two prime suspects, their threat tended to cancel each other out. He felt sure that they were not in any plot together. If one of them truly was the culprit, the other would form that secure triumvirate he and Hadley had agreed upon.

Now that he had completed his preliminary survey of Europa's fierce, rare geysers, he was open to something else. Perhaps she was thinking what he was—how do we look at the geyser data in light of possible biological processes? He had suggested this to her before, and now she might just be ready to team up in this direction. There would be time when he got back home to sift through the data, to write papers and develop new theories. If "something else" meant collaborating with a biologist on said geysers, all the better. Amanda was a lot like him. She couldn't resist a new field of research. When an opportunity arose, she would grab for it, just as he would. Now was the opportunity.

He entered the darkened dome. The cherry red sub hung from its sling, ready for its next deployment. Several small ROVs stood at the ready for the midnight descent. But as he crossed to the other side, toward the storage area, it dawned on him, surprisingly, that everybody was back in the lab.

Unless, of course, Amanda was not after new research at all. What if she was not the dainty grandmother of Hadley's perceptions? What if she was following a pattern, getting Gibson alone, separate from the others? This had been a really bad idea.

For the first time in the mission, from some primitive region buried deep in the shadows of his lower brain, down where the most profound phobias and primordial terrors hid, he could feel a current spreading down into his chest, across his back. The hair stood on his arms and the nape of his neck. It was a wave of absolute horror, threatening to break on the shores of what little sanity he had left. It was the kind of feeling that sent an explosion through every nerve ending, that sapped strength from muscle and drained every last bit of energy to the point of complete paralysis. But he would not let it.

He turned on his heels to head back to the light, back to where other people were, when it hit him. Literally. He wasn't sure what it was, but the force collapsed his leg. He slammed to the floor. He reached down to his knee. He felt warmth, moisture. Blood. This was no poisoned dart or subtle weapon. This had come from an old-fashioned silent gun.

A gun. In a pressure dome. How stupid. How desperate.

He sat up. The floor spun wildly. He took in a breath to make the stars go away. He could not hear the next blast; these kinds of guns didn't make a lot of noise. Instead, he saw the crate next to him shatter, splintering across the floor and him. The fragments scattered to his right. He rolled left, behind a storage container.

"Gibson?" The voice didn't belong to Amanda. It came from behind. It was Hadley. "You in here?"

"Hadley, get down!"

At that moment, he heard a little noise, an electronic beep, coming from the other direction, from the darkness to his right. And after the faint tone, something at the far end of the habitat assemblies exploded.

Somebody in the lab called "Fire!" The automatic doors sealed with a loud clang. Hadley was trapped inside the dome with Gibson and the shooter.

Another shot fractured the doorjamb, but Hadley wasn't there anymore. Where was she?

"What have you gotten us into?" Hadley hissed from behind. She grabbed his shoulder. He grimaced. "You're hurt!"

He nodded toward the darkness. "It's Amanda."

"I know. You were right: I'm pretty sure she's one of the Seven Sisters."

"She's over there, and she's probably got enough ammunition to take us out ten times over."

"She's firing off *rounds* in a *pressure dome?*"

Another volley powdered the ice floor nearby.

"And she knows right where we are."

Hadley muttered, "Maybe she'll run out."

Gibson leaned in urgently. "Look, she killed 5 million people. She ought to be getting pretty good at it by now." He was starting to shiver.

Hadley pulled off her boot, yanked the sock from her foot, and tied it around Gibson's knee. Slipping her boot back on, she leaned close and whispered, "That should be better. They say it slows way down after a few quarts."

Gibson muttered, "You're a regular Florence Nightingale."

"She can't keep this up," Hadley said, her tone uncertain.

"Don't you see? With us gone, there doesn't need to be any more killing. All the deaths are just a bunch of tragedies that come with exploration. Just like the last expedition."

"How can she get away with it?" Desperation permeated her voice.

"She's been getting away with it ever since the war camps. Covering it up. Slipping away. She's clever. I'm sure she's thought of something."

"We've got to do something," Hadley rasped. "She's cut us off from everybody, and they're off having that fire drill our colleagues are holding."

Hadley glanced down at her wrist. "Have you got your wristcomm on you?"

Gibson held up his arm to display the wrist monitor.

She smiled. "Can you make it inside this container?"

"Inside?"

"It's big enough. You can fit."

Gibson realized what she was up to. He gave his wristcomm to her. Hadley opened the channel on Gibson's comm and gently tossed it a few feet away.

Gibson shimmied into the box, but it wasn't empty. It already had an inhabitant who had been dead for some time.

IV

Two for the price of one, Amanda thought. This was working out quite nicely. Soon, she would be free of the past. The fire would serve as a catalyst; she could convince the others to carry out a quick evacuation with the few survivors who were left. No one would have time to examine the bodies of Hadley Nobile and Gibson Van Clive. After all the clandestine hiding, the years of subterfuge, the fleeing and maneuvering and lies, she would be emancipated. Of course, lies had a way of becoming a lifestyle, but she never got used to looking over her shoulder. With the loss of Erin, she would be more mobile, free to go where she wanted, not where it was best for *two*.

She fired again, just to keep their attention. She fired low, making sure not to hit the dome's pressure envelope, although it was Kevlica, probably rated high enough that a bullet would not penetrate. She made her way, slowly and soundlessly, across the room. She had a clear view of the crate that must be concealing Gibson. She couldn't see Hadley, but she would tend to her after.

She paused and listened. At first, there was no sound at all. Then she heard the subtlest of whispers.

"Gibson, stay here. Don't move."

Now she had them. The voice was unmistakable, and coming just from the end of the big crate. She held her gun steady, out in front, and wheeled around to the sound of the whispers. There, on the ground, lay a wristcomm.

—*—

Hadley didn't wait. As soon as Amanda turned toward the wristcomm—and away from her—she clocked Amanda, hard. The coffee mug was all she could find under the circumstances. Fortunately, Jeff Cann was not much for cleaning up after himself, or she wouldn't have had any weapon at all.

The mug broke over Amanda's silvery hair. The murderer stumbled, but she didn't fall. The hatch to the habitat area unsealed, and Jeff Cann stepped in, carrying a fire extinguisher. Tefia and Felicia rushed in behind him.

"Anybody home?" Jeff called out.

Hadley turned toward the attacker, but Amanda ran. Why? She still had the gun. Perhaps she was, in fact, out of ammunition. Maybe the gun had jammed. Maybe she knew that she couldn't get away with anything before all these witnesses.

Another sharp pop echoed across the dome. Hadley flinched, but there was no bullet impact nearby. She knew where the shot had gone. Sprinting across the dome, she found Amanda sprawled on the floor, a ragged circular wound in her temple.

The seventh of the Sisters was gone, at last.

V

"All right, I've given the wound an hour to settle down. I think you're ready for my final fix." Tefia clamped a cast-like form over Gibson's leg and flipped a switch. A quiet hum issued from the cuff, accompanied by a soft blue glow issuing from its edges.

"Better?"

Gibson nodded, but he was still clenching his jaw.

"It'll really take that pain down in a few minutes."

"Sterling?" Gibson asked through his teeth.

"I'd say he died minutes before you came into the area. Jeff had just gone out to look for him, for help with the ROVs. He couldn't have been dead more than a few minutes."

"So sad," Hadley mumbled. "All of it."

"You knew, didn't you?"

Hadley froze. She felt Gibson's eyes on her.

"You knew it all along. The connection between Amanda and your father."

"Suspected," she corrected. "The years have taken a toll on us all, and Amanda certainly didn't look like the infamous Sister she was back in the camps."

"But you knew about Orri. And Ted."

She looked toward the wall, gathering her thoughts. "Orri was part of the machine that drove the Eastern Alliance. He worked in an asteroid mine where hundreds of prisoners lost their lives. And he got away with it and didn't regret it for a moment."

"How do you know?"

"I had an opportunity to ask him."

"Did you kill him?"

'No, absolutely not. I wasn't out to kill anybody. I was out to bring them to justice, to cull information, evidence."

"And Ted?"

Hadley paused. She dropped her chin to her chest for a moment, then locked eyes with Gibson again. "Ted was a sad case. I'm sure he started out sincerely, but at some point, he sold his soul to the devil. He was involved up to his eyeballs in support of the camps."

"But everybody's guilty of something. You can't just go around all the corners of the worlds, trying to—"

She leaned into his face, her eyes desperate. "Gibson, he knew my father. He'd seen him go through the gauntlet. He could have done something to get him out, to get a lot of people out. But he didn't have the guts." Before he could ask, Hadley said, "Poor Dakota was another thing. She was innocent as the driven snow. Collateral damage, I suppose. I feel horrible…"

There were so many things Gibson could ask her. Perhaps he was too good of a friend to ask, "Did you sacrifice anything on the expedition side just to hunt down war criminals?" "Did you think you could be endangering lives by hiding the fact that you might be hiring a couple of murderers onto our crew?" To Hadley's relief, he did not. Instead, he said, "So when you put out the request for proposal for our expedition, did you write, 'Wanted: microbiologist. Must have six years experience in deep space and four years medical experimentation at Antarctic war camps.' Was that it?" He sounded hurt beyond his physical wounds.

"Gibson, I—"

Sensing an awkward moment, Tefia left the room.

"But why didn't you tell me? At least about Amanda? When the murders began?"

She put her hand on his. "Simple. If something went wrong, if I was wrong or if I was right but she got to me first, I didn't want her coming after you. But in the end, she came after all of us."

"I would have rather known so I could see her coming."

"And if I was wrong about who was behind it all, I figured you'd be caught looking the wrong way. No, I needed you to have a different perspective, a view not contaminated by the scant knowledge I had. And it was scant. Just tidbits and suspicions. On paper, it looked like I was paranoid."

He coughed. "Fair enough." Gibson held Hadley's gaze. "Are you free of the past now? Is that what this was all about?"

Hadley looked down. "Only part of it. The science is my first love. But when I got a whiff of the fact that Amanda might have been one of the Seven Sisters, I began to put a pattern together. So many people who were guilty of so many atrocities ended up in our field of work. Ironic. The science always drove me, but in the end, it was also about justice for those who got away."

"But whose justice?" Gibson asked, gently.

"Not mine. I wasn't setting myself up as judge and jury. I was just going to get enough information to set the record straight, to bring them to justice before the interplanetary court."

She looked back up at him, and she saw it in his face. Gibson was too good at reading her, and he knew she had unspoken, unfinished business. She turned her face away from him, toward the window. "Maybe I'll find the peace I need out there. At that little cache you found east of the lost expedition's dome. And we're reaching our limit on rads. No time to go fishing again. We need to go. Just one more set of loose ends to tie up."

"Our lost expedition?" Gibson asked.

Hadley nodded, gazing ahead in silence.

16

A Detour

I

"We're headed east, right?" Felicia turned to face Tefia, sitting across the glider's aisle. "Where are we going?"

"Nobody told you?"

"Nobody tells me anything. New kid on the block and all that."

"Sorry. It's all been a bit scattered, and I can't remember who I told what. We're taking a little detour on our way back home. For history's sake."

Felicia stared at Tefia, palms up toward the ceiling.

Tefia shrugged. "I'm not so sure about the details. Hadley and Gibson have something they want to check out; something about the first expedition. There's an old cache or something out there."

"It's not far," Hadley called back from the cockpit. "At least not in this thing," she said aside to Gibson. With him in the driver's seat and Hadley riding shotgun, she felt like mom and dad taking the kids on a family outing. But the family had shrunken substantially. Its only inhabitants filled the passenger cabin: Tefia and Felicia sitting across the aisle from each other, Jeff Cann keeping to himself, staring out the window, his mind a few hundred miles to the north where Dakota lay in Taliesin's medlab.

Hadley added, "You kids play nice back there. We *will* pull this glider over."

As the glider gained distance from the dome, something in Hadley envied Dakota. She lay ensconced back in the warmth and safety of Taliesin. Then again, the medics probably had the plucky marine biologist hooked up to tubes and monitors, not such an enviable position. Hadley found herself crossing her fingers, hoping Dakota would pull through her coma.

© Springer International Publishing Switzerland 2017
M. Carroll, *Europa's Lost Expedition*, Science and Fiction,
DOI 10.1007/978-3-319-43159-8_16

She soon lost any sense of envy as the glaring landscape of Europa worked its way into her heart. They had made it to a point roughly intersecting the route taken by members of the lost expedition all those years ago. Their course would take them over territory seen only by those past explorers, past the ragged ice piers and snaking floes ahead. And beyond that horizon, answers to mysteries awaited. Would they discover the ultimate fates of Culpepper and Ramirez?

Finally, the hulking globe of Jupiter appeared to rise at the horizon as the glider made its progress eastward. The clouded sphere swelled, spreading along the horizon, eating away at the relentless black of the sky.

"What's that?" Hadley pointed out the windshield. "See it, by the low shelf?"

"Got it," Gibson said, slowing the craft.

As they approached, a flash of orange and silver came into view. Clearly, it was no natural formation. Sunlight flared along a chrome line. Just beneath it, two black arcs rested against the ice.

Hadley leaned toward the glass. "Whatever it is, it's artificial."

"I know what it is," Gibson said. "Think back to the diary. Remember what Culpepper said Donnie Ramirez made?"

"A sort of sidecar ... "

"Yes, for the mini rover. Let's check it out."

The glider came to rest a dozen meters from the object. A bank of brackish material had oozed from the ground, partially burying the decade-old structure in an icy froth, camouflaging it from any orbital imagery. Tefia walked around the far side as Hadley and Gibson marveled at the thrown-together contraption. Felicia and Jeff Cann approached more slowly, letting the others take the lead.

Felicia said to him, "I'm just trying to avoid looking at the roof while we're out here."

Of course, they had no choice. The bodies couldn't be kept inside. The warm compartment would cause them to decompose, contaminating the precious oxygen reserves. Better to keep the victims outside, cryogenically frozen to the top of the glider.

"Yep. I've had enough of Halloween scenes for this trip," Jeff agreed.

"Then you won't like what I just found," Tefia said, grimacing at the back of the sidecar.

Hadley and Gibson dashed to her side. There, leering from the ice, lay a partially exposed suited figure. The grinning face inside was nearly unrecognizable, a mask of beef jerky freeze-dried over the course of more than ten years. A halo of wispy tendrils drifted out from the jawline and chin.

Hadley could hear the blood pulsing through her head, throbbing in her temples. She leaned close over the figure. "That's him."

"Which one?" Gibson asked.

"It's Dave Culpepper. He was the one with the beard. Even in those last recordings, Ramirez was clean-shaven, relatively."

"Which leaves us with Donnie Ramirez. Where could he have gone?"

"Maybe to the east, to their Science Station B."

"You guys should see this," Tefia said. She stood up from examining the figure and pointed to the back of the head.

Hadley studied the helmet.

"It appears that Dave Culpepper did not freeze to death or run out of oxygen or even expire from radiation poisoning." She pointed. Along the back of the helmet, a sharp fissure tore across the metal shell. Even now, brownish rivulets branched away from the opening, draping the helmet in a flash-frozen forest of leafy stains.

"Murder," Gibson said. "Somebody hit him. Hard."

"Let's find what's left of Donnie Ramirez," Hadley said, heading for the glider.

"What about Dr. Culpepper's body?" Tefia called.

Hadley didn't break stride. Culpepper's murder seemed to energize her into action. "No room. I wish there was, out of respect for him and his loved ones. But it's just not practical; three on the roof is enough."

II

The infamous eastern Science Station B lay just ahead, somewhere over the undulating ridges and valleys of ice. The glider's course had carried them to a point closer to Jupiter. The banded orb now displayed over half of its face, towering above the horizon. Gibson watched the glider's nav system, slaved to the orbital imagery, as the kilometers clicked down to zero. There were no global positioning satellites available in Europa's small constellation to help with the journey, although one would clear the horizon soon. He surveyed the landscape in front and to the sides.

"According to this thing, we're there."

"It's right," Hadley said, pointing out the windshield. On a slight rise, in the middle of the relentless blue and white of Jupiter's ice moon, stood the bulbous orange sections of the cache, gilded with a light frost.

"What have we here?" Jeff Cann asked dramatically, leaning into the cockpit.

"Ten-year-old supplies," Gibson said. "And maybe some answers."

"And maybe some*body*," Hadley added, emphasizing the last syllable.

"I hope they have Twinkies. They've got a shelf life twice that."

"The first one we find is yours," Hadley promised.

Gibson brought the glider to a standstill alongside the assemblage. As the lock cycled, Hadley could feel a building sense of dread. In the last few weeks, she had seen enough corpses for a lifetime.

The cache consisted of three low, slightly domed cylinders, each about four feet tall and five across. The sections reminded Hadley of orange mushrooms. The three segments met at a central hub. Each section had a port facing outward, away from the center, for access to the supplies inside. It was a large cache, but it had no tanks or pressure vessels.

"Strictly for storage in a vacuum," Gibson grumbled. "I was hoping..."

"We'll have a look around," Hadley said, making her way toward the opposite side. "Oh," she exclaimed.

Gibson and Felicia reached Hadley's location in moments. Scattered across the ice at the foot of the far section, a trail of equipment lay in piles.

Hadley picked up a cylinder and dropped it again. "Looks like a bomb went off. They were looking for something."

"They were trying to fix their rover. I'd say they were pretty desperate."

"And they never found anything. Sad."

They continued around the cache sections. It was the same story: pieces of hardware and electronics juxtaposed upon each other, strewn over the ice. One of the sections was nearly empty, an orange shell with a few cartons inside. The ten-year-old rations and tools might have made fine additions to any museum, but they would remain here, cocooned in the cold vacuum of Jupiter's environs. Gibson knelt to examine some of the artifacts.

"Looks like nobody's home," Tefia said quietly.

Hadley planted her feet firmly between two piles of discarded hardware. "And that leaves us with my burning question—just what happened to Donnie Ramirez?"

Gibson stood. "I have another burning question. What's just over that little ridge?"

"Is that what you were talking about? Something in the orbital views?"

He nodded slowly, a smile spreading across his face. "Yes. Remember the diary entry that said how construction was progressing on some 'little project to the east'? That's what I found in those orbital shots. Now's our chance. Want to come?"

She glanced at the time and pushed the tiredness from her voice. "Sure."

"Open invitation?" Jeff Cann asked.

"The more, the merrier," Gibson said. "If this is what I think it is, we'll all want to see it before we leave for home. Let's not forget where we parked."

The small troupe of explorers made their way down a hollow. As they crested the other side, Jeff arrived first. He stopped dead in his tracks, looking off into the distance.

"What—I don't understand . . . "

Tefia arrived next. "What are those?"

Felicia came up beside them. "Hey, it looks like a Europan Stonehenge!"

The others made it to the little summit, their breathing labored.

"Yes," Gibson said. "You're right, Felicia. In a sense."

Hadley surveyed the surreal landscape. Across the flat top of a nearby mesa, someone had arranged pillars of ice in a semicircle. Beyond them, Jupiter inscribed its circle of light against the sky, silhouetting the uprights against its face. On the close side of the arc of ice monoliths stood a single tapered shaft as tall as a person.

"It's a sort of sundial," Gibson explained. "I figured it out as soon as I looked at a map. This cache is not only farther into the Jupiter-facing hemisphere than the lost expedition's dome was; it's also farther north. Jupiter will always appear where it is, right there on the horizon. Sure, it goes through phases, crescent to waxing to full and so forth, but in relation to our horizon, it stays roughly in the same spot."

Jeff Cann held up a finger, as if trying to figure out which direction the wind was blowing. "Almost, but Europa librates."

"It what?" asked Tefia.

"Don't you medical types read? It librates, meaning it wiggles back and forth just slightly as it moves around Jupiter. So that means that from here, Jupiter will seem to bob around just a little."

"That's exactly it, Jeff. That's why they put this thing right here. They needed a remote science station, but when they got here, or maybe before, somebody figured out that you could use that libration of the planet to tell time."

"Like a gigantic sundial," Hadley said, awe in her voice.

"Exactly! You can tell by how many of these monoliths extend across the face of Jupiter. The amount of horizon that Jupiter crosses gives you an approximate time."

"Not to sound too pedantic," said Felicia, "but I'm sure they all had clocks on their monitors and heads-up displays. Why bother?"

"Because it's elegant," Tefia said.

Felicia frowned through her visor. "Elegance is one thing, but I got the impression these guys were hanging on by their fingernails."

"Not at the beginning," Gibson said. "And Donnie Ramirez was quite creative in the arts. They undoubtedly began it early, and they probably didn't plan on spending too much time on it."

Jeff Cann added, "Sometimes people do something just because it's clever, or new, or beautiful."

"Yes," Hadley said slowly. "From a moment when the road before them was filled with promise and they had the energy and vitality enough to build something impractical, something beautiful, something for the ages."

Gibson took in the panorama of Jupiter beyond the little faux Stonehenge. Europahenge. "And maybe in the end, after all the failures and disappointments, it's all the legacy they had left to offer."

"Perhaps it would do us all good to pause now and then," Hadley said. "Gibs, you forced me to look, really look, at the geysers. Not to study—we'd done that—but to enjoy. How was it that Ted put it? 'We're going out into the great unknown, into someone's creation.' And he declared it a beautiful thing. 'To know a creation is to know its creator.' That's how he saw it, anyway."

"I'm not so sure about that, but Steven Hawking likened science to searching for the mind of God. Maybe Ted had something there." Gibson frowned. "But I really thought we'd find Donnie Ramirez somewhere out here."

"He's not here," Hadley said with finality. "I know where he is. It's so implausible that we never considered it, but I know now. The key to all mysteries about the lost expedition isn't at the eastern cache or the southern ruins of the drill site; it's been waiting for us back at Taliesin."

III

The glider sailed northward. Everyone was silent in light of all they had seen at the eastern cache. Despite her studies, despite the images and robotic recon and eyewitness reports she had studied, nothing could prepare a person for the wilderness that was Europa, Hadley mused. Not really. No other world could boast the towering walls of the double ridges, snaking into the snowy distance. No one could describe adequately the wild variety of subdued colors, the blues and greens and mauves and tans hiding just under the snowy skin, revealing themselves in the cliff faces or the undersides of boulders. The mere fact that an ocean lay underneath puzzled early researchers, whose simple calculations showed that an ice crust would eventually freeze an ocean solid. Europa had surprised them. The ice turned out to behave in different, alien ways more than the early models predicted. And another thing contributed to the liquid ocean: simple salt. Europa's ocean was more briny than the Dead Sea. Saltwater could remain liquid at far lower temperatures. Its combination with acidic ices manufactured from Jupiter's radiation made the crust/ocean interface more pliable. The mix enabled the moon's surface to buckle and fault under the gravity of Jupiter and the other Galileans. The elegance of it all drew Hadley here in the first place. And now, they knew that life thrived down there, beneath the frozen wilderness. Dakota and Amanda had teased out the fingerprints of life, and those microbial colonies on the sea floor sealed the deal, alien spheres swaying and rolling in the volcanic heat. Jeff Cann's great leviathan was icing on the cake, a gift from Europa to them for all their trouble? The fact enthralled her: to know that another world contained dancing, swimming, rushing life as complex as anything on Earth. But a lot had happened. Now, always on her mind, always tainting the soaring crystalline promontories and the chiaroscuro ice ripples, always before her lay Taliesin. And what was waiting back there.

"Chiaroscuro," she mumbled.

"Hm?" Gibson leaned toward her.

"Chiaroscuro. Light out of darkness. I like the word. It fits so well here."

Gibson frowned. "I always thought of chiaroscuro as old white guys looming out of some dark, tea-colored painting."

"Sure, but it was more general, a Renaissance thing. Like a candle in a darkened room with just a rosebud illuminated. Or a moonbeam shining on the face of a lover by a pond. That sort of thing."

"Chiaroscuro." Gibson let the word roll around in his mouth, as if it was a new and not displeasing taste. "Light from darkness. I like it."

Hadley pointed out the window at the glistening ice against a black sky, against deep purple shadows. "Light out of darkness. There it is."

17

Taliesin

I

The expedition returned to a heroes' welcome. Handcrafted signs greeted them at the main airlock, along with champagne (where did that come from?) and music. The main galley offered a cake. But as soon as the party broke up, and as soon as they had all checked on Dakota's health—remarkably improved— Hadley made her way quietly to the storage warehouse. Sure enough, a mini rover bike slumbered there, beneath piles of detritus and old equipment. The little piece of history might have gone unnoticed, had she not been searching for it. With that confirmation, she embarked on a search for a guy with hyper-mop. She found him back in the galley, cleaning up after her team's partying. The floor glistened with mop water, and the tables shone. No one else was there except someone in the food service area.

As she approached the man, he said, "You all had a nice party, Dr. Nobile, isn't it?"

"Buy you a coffee?" she flipped her hand toward a table.

He leaned on his hypermop. It hummed in anticipation of the resumption of its cleaning. "I get my coffees free here."

"Buy me one, then," she smiled. "Cappuccino?"

"How Italian of you." He called over the counter. "A cappuccino and a latte, Billy. You know how I like mine."

Hadley sat across the table from him, peering over her cup.

"Got something on your mind?" he asked after an awkward moment.

She glanced at the nametag on his uniform. "Yes I do, *Philippe*." She emphasized his name. "Do you speak French?"

© Springer International Publishing Switzerland 2017
M. Carroll, *Europa's Lost Expedition*, Science and Fiction,
DOI 10.1007/978-3-319-43159-8_17

He wasn't drinking now. He was staring. "Not a word. Oh, I toss around the occasional déjà vu and omelette, and I'm partial to French fries, but that's about it. Why do you ask?"

"Because I don't think your name is Philippe. I think I know who you are. I learned a little trick, just recently. No matter how much a person changes, the voice is the same. I know your voice."

He cocked his head to one side. "Do you, now, *mon cher*? See? The language is coming back to me."

"You see, I remember voices. It used to be my job, back in the day, when Taliesin was just a little asterisk on the maps and we ran the show from Ganymede. Yes, I remember your voice, Donnie."

A look of panic flashed across his face.

Hadley leaned over and patted his shoulder. "You seem to want to keep your ghosts, your new identity. That's fine; your secret's safe with me."

His expression relaxed. He gazed at her curiously. "I don't think I remember you."

"I was filling in. My name doesn't even show up on some of the rosters. Somebody called in sick and they made me Capcom for a day. I don't usually share my experience." *Even with my closest friends*, she thought, Gibson in mind.

He closed his eyes and rubbed the back of his neck. "It's been so long now, it probably doesn't matter."

"Hadley Nobile, if it makes any difference." She shook his hand. "Why hide all this time?"

"Hey, Capcom, did you find Culpepper out there?"

She felt the realization cross her face before she had the chance to hide it.

"Yes," Donnie said slowly. "Saddest moment of my life. Our O_2 was low, his was pretty much out, and he was spent. Couldn't move his legs. And at that point we figured out that the sidecar thing wasn't going to work. Too much extra weight for that little minirover. But he didn't want to die alone and he didn't want me to stay with him and watch him go out in some agonizing drama. He asked me." Ramirez looked down. "Asked me to put him out of it. That's the way he said it." His voice cracked. "To put him out of it. Like some animal. So I did it the only way I could figure, with a little geology hammer we had in the toolkit."

"Why not just bleed the air from his suit, let him go to sleep?"

"He wouldn't let me. Said he wanted me to do it fast, and without him looking at me. He didn't want to see it coming. But he wasn't cowardly. I'd not say that. It was just the last wish of a dying man, and for that, I was happy to do it.

"But then you went all the way back down to the dome?"

"It was a whole lot closer than trying to make it home, and the rover battery was just about shot. But I remembered we had two others like it down at the dome, in storage for one of the little ROVs. Worth the trip, as it turned out. So I took that diary of Dave's and I added the bit about me deciding to stay behind, noble self-sacrifice and all that. I should be a writer, don't you think? It was done well. Fooled everyone."

"And you came back without anyone knowing? Didn't anybody try to get in contact?"

"Radio was one of the first systems to go down. First of many."

"But why keep your arrival at Taliesin a secret?"

"After all we'd been through I wasn't thinking things through too well. I suppose I was running paranoid. If the so-called rescue party had found Dave, they'd be looking for a murderer, and I was the only suspect around. Couldn't risk it. It was easy to sneak back in. In those days, there were piles of people pouring in and out of the place with the new construction. I just followed a bunch of newbies in and laid low until I got better. I thought I'd be dead from radiation, but, well, it made me the man I am today." He nodded at his shriveled arm.

Hadley shook her head in wonder. "And you made it all the way back on that modified bike?"

"The minirover's manual said it would never make it that far, so Dave never thought to even try. Dave Culpepper was thinking like an engineer. I was thinking like an optimist. A desperate optimist."

Hadley looked into his sunken eyes. That mercy killing was a long time ago. It didn't hold the sinister essence of the crimes carried out by Amanda and Erin—a.k.a. Aaron—or the crimes of omission caused by Ted and Orri. She felt a touch of moisture at the corners of her eyes. "I am so sorry."

"It was a risky expedition. We knew the risks." He said it flatly, a simple matter of fact.

She would have none of it. "No, I mean I'm sorry for what I put you through."

Donnie looked into her eyes, searching, squinting from one to the other, furrowing his brow in concentration. Then, the wrinkles in his forehead slackened, his eyes relaxed. His eyebrows twitched up.

"You wanted some kind of redemption. Is that what this is? It's not going to happen, Ms. Nobile. It's not needed."

"What do you mean, not *needed*?" she said, hurt in her tone.

"The Comms chief isn't responsible, right? You were just the mouthpiece for the team."

"No, I had decisions to make. What if I had done things differently? What if I could have saved those people, pulled them out before it was too late?"

"Hey, Capcom, even if you had been holding our hands, there was nothing you could do. Nothing anyone could do. I'm just glad you all made it out alive this time around." He smiled wide and patted the back of her hand. In that moment, he could have been her grand Uncle Nobile, the old Italian who was known for his kind words in the turmoil.

She put her other hand on top of his and leaned forward. "Thank you, Donnie. Thank you."

He shrugged as if it was nothing. He stood and slowly hobbled down the corridor.

Perhaps it really was nothing to him. Maybe her role had been less pivotal, less accountable, than she had feared all along. But now, after a decade, was she ready to shrug it off? To shrug off all the weight and the guilt and the second-guessing?

II

"Remarkable," Gibson said. "After all they had been through, Ramirez just kept it all to himself for all these years."

Hadley sat facing him, the two little chairs in front of the monitor in Gibson's room. They sat close, and spoke in quiet tones.

"I'm just wondering about moral duty," she said. "Should we be telling someone?"

"About what? It could be argued that Ramirez committed no crime, but carried out the last request of a dying man. I suppose he could be locked up for a few weeks on charges of forgery, false identity, those kinds of things."

Hadley shrugged. "But what would be the purpose?"

"My point exactly. The poor guy should get a pass. Heck, he's a hero, a survivor. Here, let me show you something."

Gibson pulled up a new window on his monitor. Hadley recognized the corrupted text of the diary.

"As near as I can figure, Dave Culpepper made this entry late, but in that period when some were still holding out hope for rescue."

Hadley scanned the fragmentary text on the screen. Gibson had highlighted several lines.

Life will never be the same again. Our expedition will cast a long shadow over our lives, a darkness to taint happiness, a ghost to haunt every joy.

Gibson tapped the screen. "Do you know why I wanted you to see this?"

"Sheer malevolence?"

"Not at all. It's because it doesn't have to be that way. Why should a ten-year-old expedition continue to have power over people's lives? You've still got this monkey on your back, this thing that's like a demented form of survivor's guilt—demented because it's half a generation removed from you, Capcom or not."

Hadley thought for a moment, and then nodded as if she had made up her mind about something. "*There was nothing anyone could do.* That's what Donnie said."

"And he was there."

"That does give him some creds."

Gibson leaned forward, his knee touching hers. "Do you believe him?"

Hadley didn't answer immediately. She closed her eyes and became quiet. She took in a long, deep breath, let it out, and let her tight shoulders drop a bit. She opened her eyes again.

"Yes. Yes, I think I do."

"I do, too." He gave her knee a gentle squeeze. The gesture said what words wouldn't.

"Donnie told me something else: since Dakota's been awake, he's been visiting her. Whenever Jeff Cann lets anyone in. Jeff's glued to her side, and she likes it, but Donnie and Dakota have become quite close friends."

Gibson shook his head, a wan smile on his lips. "After their yelling match in the galley, who would have guessed?" He looked at her. "And what of our yelling match?"

She leveled a stare at him. Blue. His eyes were a remarkable blue. Not the cold, sterile blue of Europan mesas, but the rich, life-filled blue of a mountain lake, of a robin's egg, of the skies of Earth. The blue of life.

Love was supposed to be the greatest, noblest of human emotions. That's what all the literary greats said, from Shakespeare to Coonts. And that's what Ted would have said, too. But for Hadley, anger had trumped all, ever since the loss of the expedition a decade ago. Since even before that, back when the mastermind criminals of The War were escaping to the ends of the Solar System. Anger was one of those tricky emotions that usually had something else behind it. Was it her guilt? Insecurity? All of the above? Everyone had hurt. Everyone suffered loss. It was part of the human condition. But she had been blinded by guilt, blind to the fact that people surrounded her who really did love her. And one of them was staring into her eyes at this moment. He had always been there, always had her back, even when he wasn't too happy with her decisions. She had pushed him away. Now, just as Gibson had made her step back and see the geysers with fresh eyes, she took a moment to see her friend in a new way, a way in which to enjoy without overthinking, to revel in just for the reveling's sake. It was even better than the geysers. She could almost feel Uncle Nobile smiling.

—*—

Gibson had always thought of Hadley as attractive, even back in high school; but there was far more to that attraction than the physical. He remembered long pauses in their correspondence as separate lives got in the way, remembered how they pinged off of each other as they traveled autonomous paths, staying in touch and meeting now and then at various seminars and conferences. But somehow, she had always been off-limits romantically, a kissing cousin or a best friend's beau.

Now, he was seeing her in a different light. She no longer held the aura of a cute younger sister; she was beautiful and brilliant and capable and, yes, even sexy. And she had led this expedition in spectacular fashion against great odds, just as he knew she would. Maybe that's what he found most attractive of all. There was a sort of darkness that she was emerging from. Chiaroscuro. And here he was, a prematurely balding guy with not much in the looks department. But gazing into those green eyes of hers, he could tell that she didn't see him that way. And as they kissed, he decided that whatever she thought she saw—or perhaps overlooked—was just fine with him.

Part II

The Science Behind the Story

18

The Science Behind the Story

Thick vs. Thin: Pizzas and Europa's Crust

Among the moons of our solar system, Europa is unique and enchanting. An elegant calligraphy etches the surface in linear and arcurate stripes. The stripes and ribbons of color painting its icy face pay tribute to forces deep within its crust, forces tied to the gravitational dances of Jupiter and its other large moons. But what do these features tell us about the structure of the glistening moon?

The heroes of our story drill through the crust of Europa to access an ocean below. Is this really possible? Perhaps. The answer depends on the thickness of the crust. Some studies propose a thick ice shell around Europa, essentially solid down to many tens of kilometers. Other models posit an ice crust just a few kilometers deep. But one thing is clear: beneath the crust lies one of the grandest oceans in the Solar System. This ocean is made possible by the gravitational taffy pull of Europa's interior, at the hands of Jupiter and its other moons. Called tidal heating, this force results in prodigious amounts of heat within Europa's core. Just as tidal heating induces volcanoes on the volcanic moon Io next door, the same forces affect Europa, to a lesser extent (Fig. 18.1).

Whether the crust is thick or thin, it appears to be decoupled, meaning that it is freely floating on a liquid water ocean, independent of the seafloor below. Evidence indicates that Europa's ocean is global, a vast subsurface sea running from pole to pole under the icy surface. As Europa's crust shifts during each Europan day, Jupiter's tidal forces trigger stress fractures, cracking the ice in great arcs as the surface turns. These arcuate formations are known as cycloids (Fig. 18.2).

© Springer International Publishing Switzerland 2017
M. Carroll, *Europa's Lost Expedition*, Science and Fiction,
DOI 10.1007/978-3-319-43159-8_18

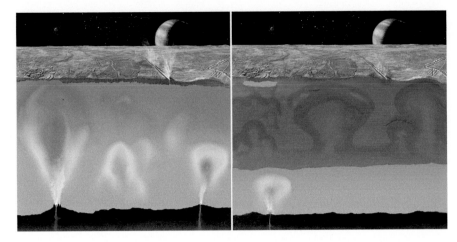

Fig. 18.1 Thick vs. thin; that is the question. Two models fit many of the features seen on the surface of Europa. One model (*left*) posits a thin crust covering a roughly 100-kilometer deep ocean. Plumes of heated water from seafloor volcanoes impinge on the base of the ice, breaking through to create the domes, bands and chaos regions. Another model suggests a thick ice crust above the ocean. Diapirs of warm ice migrate up through the ice, resulting in the surface features we see today. Chaos regions in this scenario might occur over subsurface lakes, something our characters refer to as "Schmidt Features." (This is a reference to the work of planetary geologist Britney Schmidt, researcher at Georgia Tech's College of Earth and Atmospheric Sciences) (NASA/JPL art by the author)

Fig. 18.2 Beautiful arcuate fractures dominate some areas of Europa, telltale signs of the moon's decoupled crust floating on an ocean. Ridges like the conspicuous one at top right develop into long, arc-shaped cycloids. The wall of this ridge stands half a kilometer above the surrounding plains (NASA/JPL/Caltech)

Scientists have pieced together a picture of Europa's interior by the way Europa's gravity affected the path of the Galileo spacecraft. In many of its circuits around Jupiter, the craft flew close by Europa. As the spacecraft sped up under Europa's gravitational pull, its signal shifted, just as the siren on a passing fire truck shifts. The varying structure of Europa's interior caused subtle changes in this "Doppler" shift, enabling scientists to chart the moon's internal structure. Doppler data from the closest flybys fits the model of a rocky interior capped by an outer shell of water 100 kilometers deep. Even a 6-kilometer crust would be a fragile film over such a massive ocean.

A second line of evidence for a massive ocean concerns a magnetic field emanating from within the little moon. As Jupiter's magnetic field sweeps across Europa, it triggers a magnetic field within the moon itself. Europa generates a magnetic field consistent with liquid salt water. In the first week of 2000, the Galileo spacecraft flew within 346 kilometers of Europa's surface. Later, its magnetometer detected a change in magnetic fields coming from Europa. These directional changes resembled those that would be generated by electrically conducting liquid within Europa's upper ice region. Unlike the electrical currents pouring from Earth, Europa's field is induced—it is created in response to Jupiter's mighty field lines. This induced field is generated by, and continually morphing in response to, Jupiter's rapidly rotating magnetic field.

The structure and thickness of the Europan crust is an incompletely understood issue, and a complex one at that. Europa's ridged surface has fractured into vast sections of ice called blocks or rafts. These rafts appear to have shifted and rotated before freezing solid again. Many of these rafts can be fit back together like a cosmic jigsaw puzzle, clearly indicating that a once continuous surface has been split up and moved around. Some researchers assert that the muddled chaos zones point to a thin, few-kilometers crust at the equator. They believe that the crust thickens to the north and south.

Other researchers conclude that chaos regions and other features have been generated more indirectly. What concerns many planetary geologists is the distance between Europa's ocean floor and the base of its icy crust. If models are correct, plume material would take weeks to months to migrate from the seafloor up through the water and ice to the surface. Additionally, chaotic regions would need to be heated for extended periods of time to explain the crustal movements we see. The solution to the distant plume problem may involve warm columns of ice, known as diapirs. In this scenario, tidal heating warms the interior of the ice layer, concentrated at its base, over a long period. The heat moves up through the ice much as the day-glow material in a lava lamp. Rather than melting completely through the ice, the process would be gradual and relentless, softening the ice enough to free the rafts for extended periods of time. These diapirs could also interact with pockets of trapped briny water, creating underground lakes within a thick ice crust. Chaotic regions need not be generated quickly, but could be the result of a long and gradual process. Additionally, impurities in the ice might help the

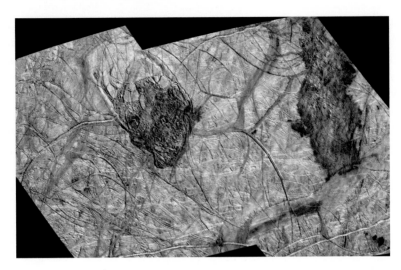

Fig. 18.3 Our character Felicia Tanaka's area of study centers upon Thera and Thrace maculae, two dark regions of bizarre terrain that pockmark the ridged plains of southern Europa. Thera, at *left*, is about 70 kilometers wide by 85 kilometers long. It is depressed below the level of the surrounding plains. Several bright rafts have broken off the edges of the area, drifting further toward Thera's interior. Its boundaries intimate that Thera formed by collapse. To Thera's *right*, Thrace Macula is marked by hummocky terrain, and appears to rise slightly above the surrounding plains. Thrace abuts the gray band Libya Linea to the south, and material from it appears to have stained the linea. One model for the formation of chaos regions is complete melt-through of Europa's icy shell from an ocean below. Other geologists contend that warm ice welled up from below in a diapir, causing partial melting of the surface

process along by the lowering of the melting point. Small amounts of salt or sulfuric acid—both of which have been tentatively identified by Galileo spacecraft data—could provide enough force to generate the domes and other features, even through tens of kilometers of ice (Fig. 18.3).

The diapir model would help to explain the lenticulae, a second class of features that may reflect seafloor volcanism. Lenticulae are dark spots or irregular areas. They come in varied forms, including apparent collapse features, sunken hollows, and upraised domes. The term is Latin for "freckles" and refers to the fact that most of these features are dark. The ruddy ice is briny, bolstering the idea that dark materials are seeping—or erupting—onto the surface from a salty internal ocean. Like the many blobs that rise in a lava lamp, these lenticulae may be the expression of the rise of warm ice from the base of the ice shell.

Amorphous, often radial discoloration may be the fallout from geyser-like activity. Many such "painted terrain" features exist on Europa. Most are associated with fractures or faults. The brown stains appear to be endogenic (generated from the inside), but whether they are localized events or global in nature remains to be determined. Some have well defined, flow-like or pooled edges. Others are diffuse, fading at the boundary. Commonly, dark material drapes the surrounding terrain, appearing more like the deposition of particulate matter rather than a flood of liquid. Just what is this dark material? Instruments aboard the Galileo spacecraft gave researchers enough data to identify several candidate substances. Investigators studying Europa's infrared (heat) data suspect various salts, with magnesium sulfate (like Epsom salts) as the best spectral match. However, other investigators argue that Europa's spectra is best matched by hydrated sulfuric acid. More recent work suggests that irradiated sodium chloride—simple table salt—has a brownish color and may be the culprit here. The inner Jovian environment is bathed in sulfur, thanks to Io's eruptions. Other scientists have suggested iron compounds that presumably issue from the rocky core. Recently, reanalyzed 1998 data from the Galileo orbiter allowed scientists to detect phyllocilicate clays. The phyllosilicates ring an area 40 kilometers wide. The broken ring lies some 120 kilometers away from the center of a 30 kilometer-diameter central crater. Research suggests that this pattern is best explained by the scattering of material from the glancing blow of a comet or asteroid. If the impactor struck the surface at a shallow angle of 45 degrees or more, some of the impactor's original material would have tumbled back to the surface. A more vertical impact would probably have vaporized it or driven its materials below the surface. Alternatively, phyllosilicates from Europa's interior might have made it to the surface through the migration of diapirs in Europa's icy crust, but this is considered less likely.

Europa's Orbit, Volcanoes and Cryovolcanism

Hadley Nobile ventures deep into Europa's sea in search of active volcanoes. Given Europa's orbital stresses, many planetary geologists consider it likely that volcanic mountains pockmark Europa's ocean floor, where the silicate core of the moon meets its liquid ocean. Core heat may well be expressed in the form of submarine volcanoes akin to those we see on the Earth's seafloor. Many of Earth's oceanic volcanoes erupt highly acidic, superheated waters. But some sites are more benign to living systems. Along the mid-Atlantic ridge, a colossal mountain range rises from the floor of the Atlantic Ocean. A long series of vents peppers the chain of summits. Known as the Lost City vents, they arise when the chemistry of seawater intermingles with the mantle rocks. Calcium carbonate grows into pillars towering some 60 meters above the seafloor. Minerals gush from the top and side fissures of the columns, releasing a mix of methane and hydrogen. As hydrogen interacts with carbon dioxide dissolved in the seawater, organic compounds form within the vent.

At other seafloor locales, we find other examples of extreme water environments: hot, high pressure water from the deep sea vents. At these vents, hot water surges from the subsurface into the ocean water. The erupting water is rich in chemicals that can sustain life. The temperature of the hottest vents is far higher than the temperature at which water usually boils (100 °C), but because of the environment's high pressure, the water remains liquid. Even in this extreme environment, these vents support diverse ecosystems. Some biologists think a similar reaction led to early life forms on Earth, and may lead to life on other worlds such as Europa, where our hero, Hadley Nobile, observed submarine vents like those at Earth's Lost City.

Estimates of Europa's ocean depth imply that those vents are submerged by as much as 120 kilometers of water. But Europa may sport two kinds of volcanism. While conventional silicate volcanoes may belch molten rock at the seafloor, cryovolcanoes may spout geysers of ice "lava"—water—from the ice crust.

It is clear that, in order to have cryovolcanism, a moon or planet must have liquid water or slush in its interior, and this liquid water or mixture of water and other materials must be able to come to the surface to erupt. A number of factors contribute to cryovolcanism, including the moon's size, the amount of radiogenic heating from its core, and the history of the moon's orbit (whether tidal heating occurs).

Another important factor in cryovolcanism is the composition of the magma, erupted lava. In the case of cryovolcanism, lava is not molten rock but rather molten ice, or water. Erupted material from cryovolcanoes is often called "cryomagma." Water by itself presents a problem, as liquid water is denser than ice. In theory, it is hard to erupt water through a less dense solid ice crust. But cryomagmas are probably not made up only of water. Materials such as salts can lower the melting point of cryomagmas and make them easier to erupt, and gases trapped within can help the cryomagma to rise. The composition of the cryomagmas will depend on the distance of the body from the Sun, and what materials condensed from the solar nebula at that distance. Rocky materials and metals tended to drift toward the inner Solar System in its formative epoch, while lighter materials accumulated in the cool outer system. Methane and ammonia likely condensed from the solar nebula at the distance of Saturn and beyond, so it is likely that these could be present in cryomagmas. While ammonia has been detected in the geyser plumes of

Saturn's moon Enceladus, it is not expected at Europa's closer proximity to the Sun. In the more distant outer reaches of the Solar System, carbon monoxide, carbon dioxide, and nitrogen may play an even greater role. But at Europa, salts and volatile gases are the most likely contributors to cryovolcanism.

The thickness (viscosity) of cryomagmas depends on its mixtures of water and other materials. Molten water would simply flood the surface, filling in topographic lows. But on Europa, we see the flows with thick margins, domed structures and bowing centers in valley floors. Slurries of briny slush may have viscosity similar to those of silicate lavas.

High-resolution images taken by the Galileo orbiter reveal a plethora of indirect evidence for ice volcanism. Glistening cracks snake across the frozen landscape, bracketed by long ridges rising hundreds of feet into the black sky, sharing similar morphology to Earth's rift zones. Halos of brown surround possible vent areas. Necklaces of bright patches align with nearby fractures, adding their features to the evidence list of possible eruptions.

A recent and more uncertain line of evidence comes from Europa observations by the Hubble Space Telescope. Ultraviolet images of Europa taken by Hubble in 2012 spotted what may have been a cloud of hydrogen and oxygen, the elements that make up water. Researchers located spikes in hydrogen and oxygen levels in two regions in Europa's southern hemisphere. Computer models show that plumes of water vapor would carry similar signatures. If the results actually point to geyser activity, the plumes of water would tower on a scale of roughly 200 kilometers high. These surges were not ongoing—like those on Saturn's geyser-ridden moon Enceladus—but rather short-lived, as they were not subsequently observed.

If the plumes are real, they may be active only when Europa is at a point near its apocenter, the farthest point in its orbit from Jupiter. As is the case for Io, Jupiter's gravity has a profound effect on Europa, subjecting it to tidal forces about 1000 times stronger than what Earth feels from our Moon.

But there is a problem: the plumes seem to have disappeared. Researchers went back over data from the Galileo spacecraft, which circled through the Jupiter system from 1995 through 2003. Their studies found no evidence of eruptions. Additionally, the NASA/ESA Cassini spacecraft, which flew by Jupiter in 2001 on its way to Saturn, also failed to detect any plume activity. Hubble observations in October 1999 and November 2012 also did not detect any geysers, but according to Southwest Research Institute's Lorenz Roth, this was to be expected. "It was clear from the beginning that this is a transient phenomenon," he said (Fig. 18.4).

In our story the rarity of Europa's hypothetical geysers is explained by the complex movements of Jupiter and the other Galilean satellites, Io, Ganymede

Fig. 18.4 Image showing the estimated location of plumes of water vapor detected over Europa's south pole in observations taken by the NASA/ESA Hubble Space Telescope in December 2012. Hubble spectroscopic data of oxygen and hydrogen in Europa's auroral emissions is superimposed over an earlier spacecraft image (NASA, ESA, and L. Roth, Southwest Research Institute and University of Cologne, Germany)

and Callisto. But while these four bodies affect the moon in varied ways at various times, their effects are miniscule. The cause and nature of Europa's fickle geysers, if they are indeed there, remains a scientific mystery.

Taliesin, Chaos and Icebergs

Taliesin Chaos does not appear on any modern maps of Europa. It is a fabrication for our story, but an informed one. A few hundred kilometers to the northwest of our novel's science base lies the Taliesin crater, located at –22° south latitude and 138° longitude. Taliesin spans a diameter of 50 kilometers, and was imaged by the Galileo spacecraft only at low resolution. It is thought to be a multi-ringed impact feature. In keeping with the IAU's nomenclature rules, the crater is named after a sixth-century Celtic poet, Taliesin, whose earliest works survive in a Middle Welsh manuscript known as the Book of Taliesin. We chose the name for our base as it typifies the Celtic naming conventions of the region.

Taliesin Crater itself borders low resolution areas of Europa that may be chaos regions. These mottled areas are associated with the huge Adonis and Sarpedon Lineae. Chaos regions display a jumble of ice blocks embedded within surfaces appearing slushy in texture. Here, ridges are seen to slip across each other in lateral faults, while remnants of ridged ground have broken and rotated, some tipping up into the airless sky hundreds of meters. Chaotic terrains commonly coexist with relatively smooth, craterless landscapes. Chaotic terrains bear a striking similarity to sea ice in arctic regions. In Earth's polar regions solid ice fractures, drifts into new positions and then freezes in place again.

Some researchers suggest that plumes of hot water, generated by seafloor volcanic sources, could thin the ice, eventually triggering a series of melt-through events. The ice would fracture, freeing rafts of surface ice to bob and rotate in the quickly solidifying lake. If the concept is correct, the rafts should rotate in the same direction as the hypothesized plume (clockwise in the north). This seems to be the case in one well-documented area, although data is limited. But this scenario demands a thin crust. Other scenarios are possible with a thicker ice crust (see "Thick vs. Thin" above) (Fig. 18.5).

Fig. 18.5 This high resolution view of Europa's Conamara Chaos region displays craters ranging in size from 30 meters to over 450 meters across. This section of Conamara Chaos lies inside a bright ray of material that was ejected by the large impact crater, Pwyll, 1000 kilometers to the south. The craters probably formed from chunks of material that were thrown out by the enormous impact which created Pwyll (NASA/JPL/Caltech)

Dozens of sites on the moon hint at past eruptive events involving water that has rapidly frozen in Europa's near-vacuum environment. One type of site lies within the "triple bands," highway-like parallel stripes that contribute to Europa's cracked eggshell appearance. The structures, called lineae, can be seen as bright ridges running down the center of a dark, well-defined band. The bands are less than 15 kilometers across, but run over Europa's face for thousands of kilometers. The stripes are directional, with bands in the northern hemisphere tending in a northwest direction, and southern bands tending toward the southwest. This directional tendency suggests a relationship with Europa's orbital stresses linked to Europa's decoupled crust. The bands are diffuse and irregular in many areas. Even the central bright medians display patchy sites with halos of bright material spilling across the dark outer band. The triple bands give the appearance of fissures leaking snows of materials of varying albedos, or brightnesses, onto the moon's glistening surface. One such band is Rhadamanthys linea, which lies across the surface of Europa like a beaded necklace.

One leading theory for triple band formation proposes a tidally induced fault breaking through the ice to the ocean below. A "cryolava" of briny water oozes up to seal the vent, while geyser-like eruptions vent from weaker locations. This style of cryovolcanism is referred to as "stress-controlled cryovolcanic eruption," and may constitute fissures penetrating all the way down to the ocean or just to warm and mobile ice below. The proposed fissure eruptions may be analogous to rift-magma eruptions on Hawaii, or to the fire fountains of Io's Tvashtar Catena.

As the region around the fracture builds vertically, the weight of the growing ridge pulls on the surrounding ice, causing parallel fractures. These cracks, in turn, develop into more parallel ridges, duplicating the process as the band widens. The triple bands may form in a different way: the ridges may mark boundaries of colliding ice plates. These compression ridges sink under their own weight. As the surrounding ice is pulled downward, the sunken troughs along the ridge fill with dark material. While the details of the triple band morphology are not yet understood, the ridges bear similarity to the mid-Atlantic ridge, where seafloor spreading builds new territory beneath the Atlantic Ocean. Several shared characteristics are remarkable. Fields of small hummocks are scattered along parallel faults, and troughs in the interior of the bands form a central axis. Terrain on either side of the central troughs appears to be spreading symmetrically away from the center.

Other planetologists suggest that linea were emplaced by solid ice rather than liquid breaking through the surface. One major argument points to the fact that many of the bands rise hundreds of meters above the surrounding plains.

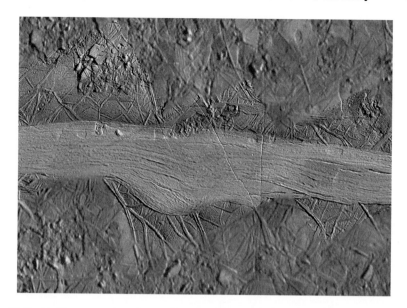

Fig. 18.6 Unlike most bands on Europa, the unusual Agenor linea is brighter than its surroundings. This Galileo spacecraft mosaic shows that Agenor is not a ridge but is relatively flat. Its interior consists of several long bands, just one of which is the very bright feature known as Agenor. Each long band shows fine striations along its length. Narrow cracks cut across Agenor, while small lenticulae dot the nearby surface. Rough chaotic terrain along the top and bottom of the view appears to be eating away at the edges of Agenor (NASA/JPL/Caltech)

This suggests ductile ice rather than liquid, which would not result in a raised structure. Additionally, there appears to be no flooding of material into adjacent ridges and valleys, which would have occurred had the linea been filled with liquid (Fig. 18.6).

The upwarped domes first pointed out by Tefia on the trip south span a diameter of roughly 10 kilometers. Fans of the volcanic plume theory point out that although long columns of heated liquid would expand and dissipate by the time they reached the ice, they may contain gases that would dissolve at the base of the ice crust. These spreading gases could be responsible for the domed structures. Many of the domes pop up in the vicinity of the chaotic regions. It has also been suggested that they result from vortices and eddies spinning off a main plume. But the ice diapir model is more consistent with the sizes of all lenticulae. Their similar sizes and spacing suggest that Europa's icy shell may be stirring like a lava lamp, with warmer ice diapirs moving upward from the bottom of the ice shell while colder ice near the surface sinks. If solid state convection is taking place within Europa's crust, warm diapirs would generate consistent scales of lenticular features (Fig. 18.7).

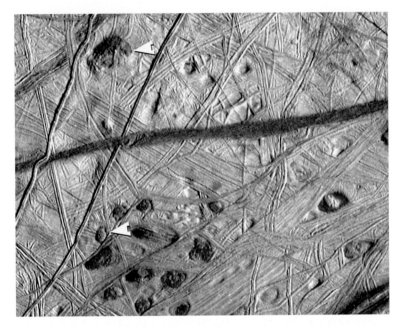

Fig. 18.7 Raised domes and sunken hollows mark the frozen landscape of Jupiter's oceanic moon. Their sizes are similar—each is roughly 10 kilometers across. Note the domes (*arrowed*) that have raised up the surrounding surface, preserving the original ridged pattern of the landscape. In this view, the surface is illuminated from the *right* (NASA/JPL/Caltech/University of Arizona/University of Colorado)

The ruddy ice seeping onto the surface may provide clues to the composition of the ocean beneath. These lenticulae might offer the best targets for surface searches for Europan life in the waters below.

Radiation and Human Exploration of the Galileans

One of the many lethal features of Jupiter's environs is radiation. The planet pours out high-energy particles in prodigious amounts. If we could see Jupiter's magnetic "bubble"—its magnetosphere—from Earth, it would appear as far across as two full Moons in the sky. The magnetic field lines emanate from the planet's core, bathing the satellites around it. Jupiter's magnetosphere acts like a giant particle accelerator, blanketing its inner moons in enough radiation to tear apart any cell walls within moments. At Io, an astronaut in a conventional space suit would receive a fatal dose of radiation in a matter of hours. The average infall of radiation on the surface measures some 540 rem/day, while the estimated total fatal dose for a human is 500 rem. But Europa receives somewhat less, and some areas on Europa are more sheltered than others. Because Jupiter rotates more quickly than Europa orbits it, the planet's magnetosphere comes up from behind, lapping the moon in its racecourse around the planet. Deadly particles constantly sweep past Europa; the most extensive radiation falls upon the trailing hemisphere. The safest places to set up camp are high latitudes on the leading hemisphere or the sub and anti Jovian high latitudes.

Europa orbits Jupiter at a distance of 676,800 kilometers, nearly twice the distance from Earth to the Moon. But the little ice ball is still well within the Jovian magnetosphere. That magnetosphere affects the surface in a variety of ways. It alters the surface chemistry as it reacts with substances in the ice. It creates oxidants, and it wears away the surface in a process called sputtering. The lion's share of erosion and radiation damage takes place in the trailing hemisphere region. There, a lethal dose will befall an astronaut in a matter of minutes. A band of high radiation also stretches across the leading hemisphere along an equatorial band. What this means for future Europa expeditions is that the safest places to go, in terms of radiation, are at high latitudes, avoiding the trailing hemisphere. In these "sheltered" regions, a lethal dose may take up to three days to build up. This would still require substantial radiation shielding, but in the time period during which our story takes place, it is reasonable to assume that such technology will be available. Our outpost at Taliesin is situated within the low radiation zone south of 50° S latitude, at about the 75° longitude line. Just 250 kilometers to the west of that location, radiation levels abruptly rise, reaching their full fury at about 250° longitude in the Dyfed Regio, at the center of the trailing hemisphere (see Fig. 18.8).

As if this deadly barrage of energy were not enough, the magnetosphere of Jupiter gets help from another source: the volcanic moon Io. Planetary geologist William McKinnon asserts that, "Jupiter is bad *because* of Io. Io pumps up the magnetosphere with all that sulfur." Io is inexorably tied to Jupiter's magnetic field. The moon's core energy is chained to the planet through a deadly super-highway of electricity called the Flux Tube. Every second, 100,000 tons of material erupts from volcanoes on Io. A large portion of that escapes into space, drifting around Jupiter in a great ring. The cloud of plasma, known as the torus, is made up of atoms of sodium and sulfur dioxide that drift away from the moon. As they spread out around Jupiter, they slam into the planet's electrical fields, becoming electrically charged (or ionized). As the particles become charged, they form the Flux Tube, a powerful sheet of electrical current. The Flux Tube sets up a sort of electrical short-circuit

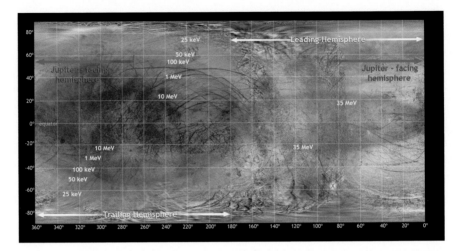

Fig. 18.8 Radiation levels vary across the face of Jupiter's moon Europa. Since Jupiter's radiation fields overtake the moon in its orbit, the most sheltered areas are north and south of the equator on the leading hemisphere. Our Taliesin Base is located at the *red* X (Art by author, based on work by Wes Patterson and Chris Paranicas, Johns Hopkins Applied Physics Laboratory; map images NASA/JPL/PDS; image processing Bjorn Jonsson, all images used with permission)

between Io and Jupiter. An astounding two trillion watts of electricity course through it, chaining the planet and moon together with power equivalent to the output of 350 nuclear power plants. As Io's volcanic activity increases, so does this cosmic link, and the entire system of Jovian magnetic fields gets pumped up to higher levels.

The biome of Earth is protected by the planet's magnetosphere, that protective bubble surrounding the globe that shunts solar radiation away from the surface. This field also helps to shield us from cosmic rays. But while Europa has an induced magnetic field from its ocean, the field is weak, and almost helpless in the face of the mighty storms of Jovian electrons.

All of this radiation sounds like a show-stopper for human exploration. It will certainly constitute one of the greatest challenges to future astronauts. For static shelters, simple ice bricks can serve as an adequate shelter (as used by our intrepid scientists atop their glider's cabin). A meter of ice creates an efficient barrier to Jupiter's radiation. But engineers continue to study other options. Although heavy, some kinds of metal can provide protection from Jovian energy fields. Several feet of water or pure hydrogen will form an effective barrier. Habitats can be designed with "storm shelters," small chambers where crews can retreat during increased radiation events, or during sleep periods. These shelters can be surrounded by fuel, or the walls can be filled with the crew's water supply.

In our story, Hadley explains to a wary Felicia Tanaka that, "The best part is that we're carrying a high yield reactor to generate energy fields around the outpost. Our dome will be shielded from most of the ambient radiation by its own little magnetosphere. It's a miniature version of what the cruise ships use on deep space trips." The use of a miniature magnetosphere for radiation shielding is a fairly new concept, but one that is gaining momentum, not only in theory but also in practical research.

The VASIMR engine mentioned in Chapter 3 generates its own magnetic field in order to focus plasma for thrust. The field could be modified to function as crew protection. The technology VASIMR uses to create its magnetic fields is superconductivity, the technology of super-conducting magnets. This is another technology that is coming of age. Superconductors can carry high levels of electric current, producing strong magnetic fields. Lightweight conductors are under development that can operate at temperatures far above those of the early superconductors, which had to be chilled down to nearly absolute zero.

A recent study done by physicists from the UK, Portugal and Sweden demonstrated that technology for generating a localized magnetosphere could be fabricated cheaply and compactly. This would enable space explorers to install such equipment on board an inhabited spacecraft in deep space, or to utilize it at the site of a remote science outpost like Taliesin. Like a planetary magnetosphere, these local fields would generate a charge in the space around the craft or habitat, shunting the deadly particles around and away from it. Using an electromagnetic probe as a model spacecraft, the team demonstrated that their magneto-ship deflected plasma around itself, clearing the space in the immediately vicinity. The inventors estimated that a spacecraft could be protected within a magnetic bubble roughly 100 to 200 meters across, using a system that could be readily carried into space. A similar field could be generated on an even larger scale for a fixed facility on the surface of a planet, asteroid or moon.

Another study, begun in January of 2013, is called SR2S (for Space Radiation Superconducting Shield). The project is being carried out by CERN and several French, Italian and international European partners. SR2S plans to engineer and validate several key enabling technologies to shield crews from radiation. Their goal is to generate protective magnetic fields using superconducting magnets.

Researchers at Johnson Space Center have been experimenting with high temperature superconducting structures to generate protective fields. Project engineers assert that, "a combined system of active and passive radiation shielding constitutes the most promising solution to this issue." JSC engineers envision large, ultra-light, expandable coils to reduce long-term radiation exposure of humans in spaceflight. A recent NASA report stated, "Ultra-light HTS coils offer significant deflection power for charged particles and due to the low amount of material from the HTS magnets, secondary particle production is kept at a low level." In the severe environment of Europa, this technology would be critical to any long-duration stays on the surface.

Biomimetic Robots

Our story's expert on human/submersible interfaces, Sterling Ewing-Rhys explains to his team how ROVs work: "It's weighted to stay upright. Designed to emulate the glass knifefish. These others are based on tuna, lamprey, salamander…nature's designs are often best."

Taking inspiration from designs found in nature, biomimetic robots, or biobots, emulate the agility and mobility of biological forms. The ocean's inhabitants exhibit high-endurance swimming that outpaces current underwater propulsion technology for stealth, maneuverability and speed. The rays, considered some of the most high-performing species, migrate over long distances with sustained speeds of 2.8 m/s. They can maneuver in tight spaces and produce enough power to actually jump out of the water. Our story's prime example of biomimetic design is the glass knifefish (or "glassfish"). This marine creature uses a single ventral fin, which runs the length of its entire body, to change direction or hover in place. The fin undulates rapidly, sometimes in counteracting waves. A team of marine engineers at the University of Edinburgh is working on a parallel design with its "SquidROV." The purpose of the project is to develop a biomimetic submarine propulsion system for a remotely operated vehicle. As Sterling asserted, such a propulsion system is more efficient than a propeller-based system of equivalent thrust. It also generates much less turbulence, making it ideal for research and observation of a host of marine creatures. The SquidROV propulsion system is based on an undulating fin system, similar to that of the glass knifefish or manta ray. The undulating fin generates a traveling wave along the length of the fin, producing thrust and maneuverability for the vehicle. The design of the vehicle utilizes ten servo motors mounted on a rigid surface with a flexible membrane mounted between them.

Another project, carried out at the Swiss Federal Institute of Technology in Zurich, tested a "nautical robot" that incorporated four fins inspired by those of the cuttlefish. Called Sepios, the 70-centimeter-long ROV exhibits a high degree of maneuverability, along with the capacity for omnidirectional movement.

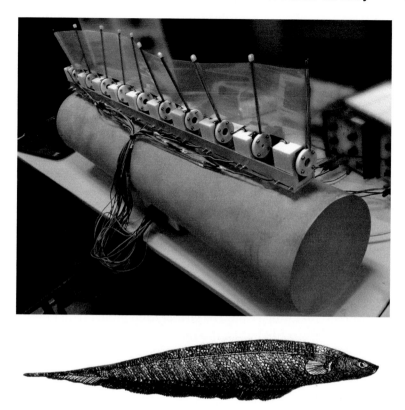

Fig. 18.9   *Top:* the biomimetic SquidROV lies upside-down on a test stand, awaiting a submarine voyage at the University of Edinburgh (photo Johan Magnussen, used by permission). *Bottom:* the glass knifefish lends its form to many experimental submersible designs, including the SquidROV. Note the single fin along the belly (Derivative work originally from Naturalist on the River Amazon, 1863; via https://commons.wikimedia.org/wiki/File:Gymnotus_carapo.png; author amended image)

Researchers at Canada's Dalhousie University have combined forces with McGill University and York University to create the AQUA robot. This aquatic vehicle is a hybrid, with the capacity to walk along the seabed as well as to "swim" using its legs. AQUA is capable of five-degrees-of-freedom motion in open water. It can also swim along the surface. The vehicle itself is augmented with a variety of sensors that can be used to calculate the AQUA's position with respect to its surrounding topography as well as a global frame of reference.

The dangers intrinsic to human exploration of a Europan ocean will undoubtedly be overcome in time, but before humans go, their emissaries will pave the way. Some of those robotic ambassadors will carry the visage of Earth's living things as they ply the alien depths of another world (Fig. 18.9).

Power Tethers

The Taliesin science outpost is powered by a clever tether system, tapping into the great energy fields pouring in from nearby Jupiter. The character Tefia Santana says, "The tether goes for 200 klicks to the west, well into the high radiation zone. As Jupiter's magnetic field sweeps across Europa, the tether gathers power and sends it to Taliesin."

A study conducted at the European Space Agency, led by Dr. Claudio Bombardelli (now at Madrid Technical University), investigated the use of tethers to generate electricity within the Jovian magnetosphere. His team's concept, which looked at power for orbiting spacecraft, was based upon the same principle as electric motors or generators. A wire moving through a magnetic field builds up an electrical current. Bombardelli's team estimated that a spacecraft in polar orbit around Jupiter could generate a few kilowatts of power using two short (~3-kilometer-length) tethers. If those tethers were lengthened to 25 kilometers, power generation could reach into the megawatt range. An outpost on Europa would be farther from Jupiter than Bombardelli's case study, which assumed a closest distance (periapsis) of less than 1.3 Jupiter radii. At Europa's distance, the system would have access to less power from Jupiter's magnetosphere. Longer tethers would be required to generate power for a facility like Taliesin, but these tethers could simply be laid along the surface of the moon like undersea cables.

In fact, NASA has conducted several experiments with electricity-generating tethers in space. The most extensive was carried out on February 25, 1996. The space shuttle *Columbia* deployed an Italian satellite at the end of a 20,000-meter tether to monitor electrical current generated by the spacecraft's movement through Earth's magnetic fields. As astronauts unrolled several kilometers of tether, the electrical current grew at the pre-dicted rate. By the 5-hour mark, the experiment was nearly completed when the tether suddenly broke, severed by the stress of extreme voltage. Engineers back at (then) Cape Kennedy studied the frayed end of the tether, and pieces of the cable were tested in a vacuum chamber. The nature of the break suggested it was not caused by excessive tension but rather that the extreme electric current had melted the tether. Tests suggest that the failure was caused by an electric arc generated by the conductive tether's movement through Earth's magnetic field. Despite its premature finale, the experiment successfully generated 3500 volts. At Jupiter, available electricity is much greater. Jupiter's deadly magnetosphere might turn out to be an asset to future visitors.

The Question of Life

Our biologist character Amanda asserts that, "… any Europan life may not be based on DNA we can recognize, but there's a theory that the DNA structure, or something similar to it, is universal to life, so there we go." If microbes do sail the seas of Europa, they may be difficult to recognize. Researchers will be looking for "handedness," the direction in which a structure like DNA tends to spiral. DNA is built like the rungs of a ladder between outer lines. Imagine taking those lines—the ends of the ladder—and twisting them. This twisting movement creates the double helix form so familiar to biologists. Terrestrial DNA is almost exclusively right-handed (right twisting) (There are rare forms of DNA that are left-handed. These are known as Z-DNA.). If left-handed DNA comes up in samples, it is likely from an alien source.

Another test is to determine if the organic structures have chirality. A chiral pattern is one that is discernible from its mirror image. If the pattern has the same "outline" but cannot be flipped to exactly imitate its mate, it has chirality. An example in nature often recognized is the human hand. If the right hand is flipped so that it is a mirror image, it does not match the left hand. Chirality is found in biological structures.

In addition to right- and left-handed DNA, biologists will hunt for complex microstructures, and for chemistries associated with known biological systems. Any chemical imbalance will be subject to scrutiny. The life-engendered oxygen in Earth's atmosphere should be sequestered in its rocks, combined with other elements, or largely lost to space. It is a biological marker, present because of the operation of life. Another marker is methane, a gas that does not hang around for long in a planetary atmosphere. Methane combines with other gases or drifts away into space unless recharged by either biological processes or volcanism.

The candidacy of Europa as a possible abode for life skyrocketed with the discovery of extremophiles, microbes that thrive in extreme environments. Hypersaline lakes, often found in inland deserts such as the Great Salt Lake, play host to halophillic archaea, microbes able to tolerate high concentrations of salt. These extremophiles survive the briny environment by actively pumping the salt out of their cells. They survive not by adapting to their extreme environment but by protecting themselves from it. Halophiles would be perfectly at home in the salty realm of a Europan sea.

Another group of microbes adapted to extreme environments are the hyperthermophilles. These specialized life forms thrive in hot springs that flow from volcanic regions. Hyperthermophilles develop specialized types of enzymes—catalysts that speed chemical reactions—that remain stable at high temperature (Fig. 18.10).

At some mid-ocean ridges 3 to 4 kilometers beneath the surface of Earth's oceans, bacteria form the basis of a deep-sea food chain of creatures such as giant tubeworms, deep-sea crabs, shrimp-like crustaceans and mussels. Hydrothermal vent organisms are influenced by deep-ocean current patterns, and by topographic features such as deep fracture zones or changes in depth. One deep-sea biome stretches across more than 30 degrees of latitude along the East Pacific Rise.

These surreal ecosystems are completely cut off from the Sun's energy, nurtured by the chemical plumes of seafloor volcanoes. As such, they may be a fine example of what biomes might be possible in the hadal waters of Jupiter's ocean moon. Perhaps one day, a future version of Hadley Nobile will dive through Europa's ice crust and report back to us, firsthand, on the alien gardens of Europa's ocean floor.

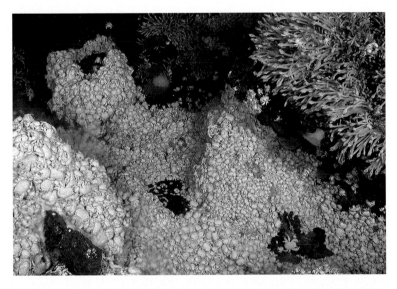

Fig. 18.10 At an ocean depth of 2397 meters, extremophile *Kiwa N.* crabs encrust a volcanic vent, while the stalked barnacles *Vulcanolepas* cling to a chimney at *right* (Photo by A. D. Rogers, et al., from *PLoS Biology*, via Wikimedia Commons, https://commons.wikimedia.org/wiki/File:Dense_mass_of_anomuran_crab_Kiwa_around_deep-sea_hydrothermal_vent.jpg)

Printed in the United States
By Bookmasters